The Last CEO

Other books written by Dave Karpinsky

Artificial Intelligence & Information Technology
- Artificial Intelligence (AI) for Daily Life: A Practical Guide to Artificial Intelligence
- AI and Creativity: How Machines are Changing Art, Music & Literature
- AI-Powered PM: Leveraging Artificial Intelligence for Enhanced Efficiency and Success
- Artificial Intelligence Rise and Humanity Fall
- Data-Driven Future: Harnessing AI and Big Data for Tomorrow's Challenges
- Deepfake Technology: The Dark Side of AI, Manipulation and Digital Deception.
- Fixing Failed Projects: How to Master the Art of Project Turnaround
- From Data to Decisions: The Role of AI in Business Intelligence
- Jobs AI Will Replace: Re-tool or Be Left Behind
- Mastering Advanced Project Management: Strategies for Excellence
- Mastering Project Management: In complex, stressful & high-pressure environments
- SAP S/4 Implementation: A Comprehensive Guide for Practitioners
- SAP S/4 Implementation Methodologies
- SAP S/4 Implementation – Volume 1: Prep & Explore Phases
- SAP S/4 Implementation – Volume 2: Realize & Deploy Phases
- SAP S/4 Implementation – Volume 3: When Projects Fail
- The Five-Day Organizational Change Manager
- The Five--Day Project Manager
- The Project Management Masterclass: Advanced Techniques for Success

- The Rise of Real-Time Analytics: Speed, Precision, and Competitive Edge

Business & Finance
- Building Wealth in Developing Nations: A Comprehensive Step-by-Step Guide to Empower Emerging Markets
- Chief Executive's (CxO) Playbook: The First 90 Days Guide to Success
- Creating a Deployment Plan: Navigating Complexity to Deliver Success
- Creating a Strategic Roadmap: Crafting the Blueprint from Vision to Execution
- Investing Strategies of the Rich and Famous: Discover How to Diversify Your Portfolio for Maximum Returns
- Outsmart the Game: Winning When the Rules Are Rigged
- The Data Delusion: Exposing False Metrics That Shape Your World
- The Last CEO: How Artificial Intelligence is Redefining Corporate Leadership
- Trust is the New Currency: How Connection Wins in the Age of AI

Life Coach & Mentor Series
- Aspiring Entrepreneurs
- Bored Housewife
- Career Transition
- Couples and Relationships
- Mid-Life Crisis
- Mindful Healthy Living
- Project Managers
- Seeking Life's Purpose
- Surviving Holidays with In-laws

Science & Physics
- Game Over. Reset Earth
- Quantum Entanglement: The God Effect and the Secrets of Reality
- Multiverse Parallel Dimensions: The Theories and Possibilities of Parallel Universes
- Space-Time Continuum: Navigating the Quantum of the Fourth Dimension
- The Hubble Tension: The Universe's Expansion, Cosmology Crisis, and the Limits of the Big Bang Theory
- The Singularity Shift: Unveiling the Future of Humanity and Intelligence
- Twin Paradox: Solving the Puzzle of Special Relativity

Sociology & Politics
- America at War: Russia, China, Iran, S Korea
- Blue Zones Volume 1: Mystery and Science of Blue Zones
- Blue Zones Volume 2: Longevity Lessons of Blue Zones
- Decline of American Supremacy: Understanding the Erosion, Shaping the Future
- Future of Military Technology Powered by AI: How countries are transforming their warfare
- Herd Instinct: Understanding the Human Psychology of Collective Behavior
- Our Idiot Species: Evolution in Reverse
- Preventing Squatters: A Comprehensive Guide to Protecting Your Property
- Puppet Masters: The Hidden Hands of Political Power
- The Great War of China vs Russia: A Future Battlefield that Reshapes the World
- The Modern Stoic: 365 Ancient Practices for Wisdom, Peace, Purpose ad Strength
- The Next Battlefield: How AI, Robotics, and Biotechnology are Transforming Warfare

- The Savage Guide to Winning: The Brutal Truth About Success
- The Trump Effect: Return to the White House
- The Vatican Murder Cover-Up
- Unf*k Yourself: A No-Bullsh!t Guide to Taking Control
- Warfare Redefined: Military Technologies and Tactics of Tomorrow's Superpowers
- Zero F*cks Given: How to Stop Worrying and Live Your Life
- God & AI Series:
  - Is There God: According to Artificial Intelligence (AI)
  - What is God: According to Artificial Intelligence (AI)
  - What is God's Plan: According to Artificial Intelligence (AI)

"The future CEO isn't a person in a corner office —
it's a system of collective intelligence."
— *Dave Karpinsky*

The Last CEO:

How Artificial Intelligence is Redefining Corporate Leadership

Dave Karpinsky, PhD, MBA, PMP, Prosci

Green Parrot Media

Contents

Preface

Picture a boardroom late at night. The table is polished, the chairs are empty, and the walls are lined with glowing screens. Markets are shifting across continents, supply chains are recalculating by the second, and customer sentiment is pulsing in real time across millions of devices. For generations, this was the stage where CEOs and executives gathered to weigh risks, debate options, and make choices that shaped the future of their companies. But imagine, in this moment, no human leader presiding over the room. The decision is not made by instinct, compromise, or charisma. It is made by a system designed to see what no person can.

The story of leadership has always been human. Kings ruled kingdoms, generals commanded armies, and chief executives rose to guide corporations. They carried the burden of judgment and the responsibility for both triumph and failure. Yet human leadership has never been free from error. Bias colors perception, ambition clouds reason, fatigue narrows judgment. The history of business is filled with examples of leaders who had access to data and expertise yet made poor decisions because personal incentives or emotions overpowered the facts.

Consider the financial crash of 2008. Executives had the numbers. Risk reports warned of instability. Analysts flagged the dangers. Still, decisions were made that ignored or downplayed the signals. Overconfidence, fear of missing out, and misaligned incentives drove choices that triggered

global consequences. Or think of Sears and Toys "R" Us —
retail giants that saw the rise of online shopping yet failed
to act. The information was there, the trend lines clear, but
leadership clung to tradition, unwilling to face the hard
truth until it was too late. These failures weren't about a lack
of data; they were about the human limits of decision-
making.

Artificial Intelligence (AI) is not burdened by those limits.
At its best, AI does not sleep, does not tire, and does not
cling to pride. It can process millions of data points in
seconds, weigh competing factors with precision, and
present recommendations free of ego. Imagine an AI that
can analyze every quarterly earnings call in the S&P 500,
extract sentiment, compare tone across industries, and flag
predictive patterns for future performance. A human
executive might review a handful of competitor reports. The
AI can review them all, instantly, without bias. Or consider
the supply chain crisis during the pandemic. While
executives scrambled with incomplete information, AI
systems were capable of simulating routes, pricing
scenarios, geopolitical risks, and shipping bottlenecks in
real time. These were executive-level decisions being made
by machines.

In truth, this is already happening. Amazon's predictive
algorithms guide what products to ship before customers
even place orders. Hedge funds rely on trading systems that
operate with speed and consistency no human trader could
match. UPS saves millions each year with AI systems that
map delivery routes, adjusting for weather, traffic, and
demand. These aren't "tools" operating on the sidelines.

They are central decision-makers acting at the highest level of strategy.

The question this book asks is not whether AI will influence leadership. That moment has already arrived. The real question is how far it will go. Could a corporation one day be run entirely without a human CEO? Could the corner office be replaced by an algorithm that reports to the board? And if so, what role remains for people?

The answer is not the end of leadership, but the birth of a new form. Leadership will shift from one person's judgment to a blend of machine precision and human meaning. An AI may decide how to allocate capital, but a human will still need to explain that decision to employees and investors. An AI may recommend closing a factory, but humans will still manage the culture, the morale, and the social impact. The strength of AI lies in clarity and scale; the strength of humans lies in empathy and connection. Together, they create a hybrid model that redefines what leadership means.

Of course, such a shift will not come quietly. Executives will resist, fearing a loss of power, status, and pay. Boards will hesitate, unsure of how to hold an algorithm accountable. Employees will question whether they can trust a system that doesn't feel, doesn't listen, doesn't understand. Regulators will debate liability when an AI makes a decision that harms. These are valid concerns, and they must be addressed. But they do not erase the opportunity.

For those who embrace change, AI-led leadership offers the promise of faster, more accurate, and more transparent decision-making. Companies that adapt will be leaner, more responsive, and more competitive. Those that resist risk falling behind, held back by human hesitation in an economy moving at machine speed. The transition is not about replacing every human trait but about redefining leadership so that bias, ego, and delay no longer hold companies back.

The Last CEO is not a prophecy of doom. It is not about machines crushing human worth. It is about seeing leadership with fresh eyes, questioning assumptions about who gets to decide, and reimagining how companies can thrive in an era when data is abundant but judgment has too often failed. It is about the partnership between human purpose and machine clarity. And it begins with a simple, unsettling, exciting possibility: **that the CEO as we know it may be the last of their kind.**

I hope you enjoy this book.

Dave Karpinsky

"Leadership once lived in a single mind. Now it lives in the flow of data."
— *Dave Karpinsky*

Part I: The End of the Human Monopoly

1: The Rise of the Algorithmic Leader

The boardroom is alive with voices trading views, yet one voice carries a different weight. It doesn't belong to a person but to a system. As discussion unfolds, it consumes financial reports, tracks global news, parses supply chain disruptions, runs simulations, and delivers a recommendation no individual at the table could match. Executives glance at one another—some skeptical, others intrigued. Dashboards and analytics have long been part of their work, but this feels unlike anything before. This is no longer advice on the margins. This is the decision itself.

For decades artificial intelligence was seen as support. It wrote customer service scripts, flagged fraud, scored credit risk, and optimized delivery routes. Companies leaned on it to reduce cost and improve efficiency. It was a tool, nothing more. That perception is already outdated. In many corporations, AI has moved from the back office into the boardroom, not only recommending but actively deciding.

The shift began quietly. Hedge funds started with algorithmic trading platforms. These platforms consumed decades of pricing data, layered it with live feeds, and executed trades in microseconds. Traders went from making bets on instinct to watching algorithms set strategy. Some firms now attribute the majority of their returns to systems, not human managers.

Airlines use AI to reroute flights across thousands of airports when storms roll in. These systems adjust staffing, gate assignments, and maintenance schedules within minutes. Human managers oversee but rarely overrule them. What began as support has become command.

The term algorithmic leader captures this transformation. It is not about a tool but about authority. When a system can decide faster, more accurately, and with less bias than a person, it becomes the executive voice in the room. This shift does not happen in isolation. It rides on trends that have been building for years: exponential growth in computing power, machine learning models trained on trillions of data points, and the explosion of corporate datasets covering every action from customer clicks to employee performance metrics.

For an experienced executive, the question is no longer whether AI can support decision-making. It is whether AI should hold the top role itself. The answer is taking shape across industries.

Consider hiring. Traditional executive teams rely on panels, interviews, and résumés, each shaped by bias. AI systems can now review thousands of candidates, cross-reference performance data, assess skills through simulations, and predict future fit with remarkable accuracy. When Unilever shifted parts of its hiring to AI-driven video assessments, the company reported shorter cycles and more diverse outcomes. This is not about saving time. It is about making choices that align more closely with performance data than with human preference.

Or consider capital allocation. Multinationals manage billions in annual investment choices. AI systems can analyze market volatility, regulatory trends, and consumer demand across hundreds of regions simultaneously. They can present optimal mixes of spending on factories, marketing, R&D, and acquisitions with a precision that surpasses human analysis. Executives who once debated for weeks can receive a clear recommendation in hours.

These examples highlight a broader reality. AI is moving up the corporate chain of command, not by force but by performance. It demonstrates competence in areas where human leaders falter: scale, speed, and consistency. When a system consistently makes stronger calls than the CEO, pressure builds to give it more authority.

The story of leadership has always been human. Charisma, vision, and influence shaped decisions. But charisma does not forecast currency swings, and vision does not rebalance global supply chains. Influence does not guarantee good judgment. As the economy becomes more complex, leaders face challenges that exceed human capacity. That is where the algorithmic leader steps in.

Executives often argue that leadership is about intuition. Yet research on behavioral economics shows intuition is riddled with bias. Daniel Kahneman and Amos Tversky demonstrated how anchoring, overconfidence, and loss aversion distort judgment even among experts. AI is not immune to bias, but when designed well, it can expose and counter human distortions. It can run experiments across thousands of variables without the emotional drag that weighs on people.

To understand how this applies, imagine an AI tasked with pricing strategy. A human pricing committee might meet monthly, scan competitor moves, and set broad guidelines. The AI adjusts daily. It monitors online purchasing patterns, tracks social sentiment, and shifts price points by geography and time of day. Retailers like Zara already rely on algorithms to manage production and inventory decisions with this kind of precision. The result is tighter alignment between demand and supply, fewer markdowns, and higher margins.

Executives who want to prepare for this future need to take practical steps today. That means moving beyond dashboards and into decision automation. Start with targeted domains where AI has proven reliability:

- Finance teams can use AI for real-time forecasting rather than quarterly models.

- Supply chain leaders can adopt AI-driven platforms that adjust routes and vendor allocations in minutes, not days.

- HR departments can deploy AI to screen résumés and predict attrition risk with stronger accuracy than surveys.

These are not experiments. They are building blocks for algorithmic leadership. By giving AI control over tactical decisions, companies gain confidence in scaling it toward strategic decisions.

The next step is governance. If AI begins making calls once reserved for executives, boards need frameworks to monitor and validate. That includes clear audit trails for algorithmic decisions, transparency into data inputs, and defined thresholds for human override. Without this structure, adoption stalls. With it, adoption accelerates. Forward-thinking firms like DBS Bank in Singapore have already formalized AI governance councils that sit alongside risk and compliance committees. This signals the seriousness with which leaders must treat algorithmic authority.

Another step is cultural preparation. Employees often fear that AI will strip them of relevance. To counter that, executives must frame AI as a partner. When ING introduced AI into its credit risk assessments, managers were trained to interpret model outputs, challenge them, and explain them to stakeholders. This preserved human roles while improving outcomes. The cultural lesson is clear: adoption works when employees see themselves as translators and stewards of AI-driven choices, not passive recipients.

The rise of the algorithmic leader is also reshaping board dynamics. Boards once measured CEOs by track record, charisma, or the ability to inspire confidence on earnings calls. With AI in the mix, boards must learn to assess systems. They must ask: how accurate is this model across time? How transparent are its inputs? What risks does it amplify? These are new competencies, and they demand directors who understand both data science and governance. This means recruitment for boards is already shifting. Firms now look for directors who can interrogate models with the same rigor they once reserved for financial statements.

Skeptics argue that no machine can replicate the instinct of a great leader. But instinct is often nothing more than pattern recognition shaped by experience. Machine learning operates on the same principle, except at an exponentially larger scale. When a CEO says "my gut tells me," what they often mean is "I have seen this pattern before." The AI sees not dozens of patterns but millions. That scale changes the game.

The transition is not abstract. It is measurable. In healthcare, AI systems already recommend treatment plans that outperform physicians in diagnosis accuracy for conditions like breast cancer. In logistics, DHL uses AI to optimize global delivery networks with adjustments made in seconds. In finance, JPMorgan Chase relies on AI to review contracts, cutting review time from thousands of hours to seconds. Each of these represents a step toward leadership functions once thought untouchable.

For executives reading this, the lesson is direct. Begin building comfort with ceding authority to systems where evidence supports it. Measure outcomes not against perfection but against the human baseline. Create forums where AI outputs are reviewed and challenged but not ignored. And most importantly, prepare yourself for the day when the question is not whether AI belongs in the room, but whether it should sit at the head of the table.

The algorithmic leader is not waiting for permission. It is already taking root in decisions about hiring, capital allocation, risk management, and strategic direction. The companies that thrive in the years ahead will be those that embrace this shift deliberately, building governance, culture, and strategy around it. Those that cling to traditional models of leadership will discover that the competition is not just another company with a sharper CEO. It is a company run by a system that sees more, acts faster, and never tires.

2: The Decision-Making Problem

A leadership team gathers at a long table, reports stacked high, analysts waiting nearby. The CEO studies a slide crowded with projections, and everyone in the room feels the stakes. The company faces pressure, and the choices made here will ripple through thousands of jobs and billions in shareholder value. Yet the decision unfolding isn't guided solely by data. It bends under the weight of personalities, the pull of politics, and the blind spots that cloud human judgment.

This is the decision-making problem. It's not that executives lack intelligence or discipline. Many are brilliant, driven, and seasoned. The problem is that human cognition comes with built-in flaws. Leaders tell themselves they are rational, but decades of research in psychology and behavioral economics shows otherwise. Bias, politics, and error creep into every corner of corporate judgment. These forces are subtle but powerful. They often outweigh facts, leaving decisions vulnerable.

Cognitive bias is the first barrier. Take confirmation bias. Executives often search for information that supports their preferred outcome while ignoring data that contradicts it. In strategy meetings, slides are chosen to back a story already in motion. Risks are acknowledged but downplayed. Overconfidence bias compounds the problem. CEOs with track records of success begin to trust their instincts more than evidence. They double down when caution is wiser. Anchoring bias is another trap. A forecast number or valuation presented early in a meeting frames the rest of the debate, even if better evidence surfaces later. These are not rare failures. They happen every day in boardrooms across industries.

Then there's politics. No decision at the executive level is free from power dynamics. Personal ambitions, turf wars, and alliances shape how information is presented and received. Executives may withhold details to protect their division. A rising leader might back a plan to curry favor with the CEO, even if the numbers don't add up. Boards sometimes approve strategies not because they believe in them, but because they fear the cost of open disagreement. Politics, by its nature, rewards short-term wins over long-term sense. The price is paid when decisions that look safe in the room collapse under the weight of reality.

Human error adds another layer. Fatigue clouds thinking during long cycles of negotiation. Emotions surge under pressure, leading to reactive calls that ignore the bigger picture. Past experience, once a strength, can turn into a trap when leaders assume the next challenge looks like the last. Even memory betrays decision-makers. Studies show recall is often skewed toward dramatic events, not accurate baselines. A CEO who remembers surviving a downturn may misread signals in the present, assuming the company is stronger than it is.

For a long time, companies accepted these flaws as part of leadership. The human element was seen as unavoidable, even necessary. After all, leadership has always been framed as art as much as science. But as markets move faster and consequences rise, tolerance for flawed decisions is shrinking. Boards and shareholders expect accuracy. Employees expect fairness. Customers expect relevance. Traditional decision-making cannot keep up with these demands.

This is where AI enters the frame. Unlike human leaders, AI does not carry ego or fatigue. It doesn't push for promotion or protect its turf. It works with scale, consistency, and clarity. That doesn't mean AI is perfect. But it is immune to many of the distortions that plague human judgment. It can analyze all the data available, not just the subset chosen for a meeting. It can test thousands of scenarios without being anchored to the first number on a slide. It doesn't forget, and it doesn't cling to past experience when the context has changed.

One striking example comes from logistics. Global shipping firms once relied on committees to adjust routes when disruptions occurred. The process was slow, political, and often wrong. AI-driven systems now adjust shipping patterns in real time. They weigh weather, tariffs, fuel prices, and capacity across hundreds of ports. The result is better reliability and lower cost. Decisions once shaped by fatigue and bargaining are now shaped by clarity and speed.

Executives can take direct lessons from this. The key is not to wait for AI to take over the entire process, but to insert it where human flaws are most damaging. Start with forecasting. Replace manual forecasts with AI-driven models that run daily or hourly updates. Then move to risk assessment. Use AI to identify hidden correlations that humans miss, such as the way supplier delays might connect with currency fluctuations or geopolitical events. Expand to pricing. Let AI suggest price adjustments across markets, testing elasticity in real time. Each of these applications removes bias and politics from areas where mistakes carry the highest cost.

Another step is to redesign the meeting itself. Too often, executives debate based on narrative first and evidence second. AI can change this. Imagine entering a meeting where the first agenda item is a summary from an AI system that has analyzed every relevant dataset. The human conversation starts from facts, not stories. That shift alone reduces the grip of confirmation bias. Leaders can still debate priorities, but they can't ignore the baseline evidence.

Boards should act here as well. Many already receive AI-driven risk assessments that flag financial and regulatory exposure. Expanding that practice is practical and necessary. Boards should demand transparency into how AI contributes to recommendations, then use that insight to hold executives accountable. For example, if AI forecasts a higher probability of supply disruption but leaders choose to ignore it, the board should ask why. Accountability expands when AI is in the room.

Of course, AI doesn't eliminate politics. It can expose it. When AI outputs contradict a leader's proposal, the tension is visible. This can make meetings uncomfortable. But discomfort is not a flaw—it's a safeguard. It forces executives to explain choices not just in terms of narrative but in terms of evidence. For companies serious about improving decision quality, this exposure is progress.

Executives looking to act can structure their organizations to support this shift. The following steps provide a framework:

- Identify high-impact decision areas where human bias has historically caused mistakes.

- Deploy AI tools to generate independent recommendations before human debate begins.

- Build governance that records both AI recommendations and final decisions for review.

- Train leaders to interpret, challenge, and communicate AI outputs with clarity.

- Create feedback loops to measure whether AI-backed decisions outperform human-only calls.

This is not theory. These steps can be implemented today. Some firms already do it in parts of their operations. The challenge is extending it into the heart of leadership, where the stakes are highest.

Scholars debate whether machines should be trusted with decisions that shape human lives. Some argue algorithms lack context and empathy. Others argue the greater risk lies in clinging to flawed human judgment when better options exist. This debate matters. But what cannot be ignored is the evidence: when designed and governed well, AI consistently reduces error and bias compared to human decision-makers.

Executives often pride themselves on decisiveness. Yet decisiveness without accuracy is simply risk in disguise. The companies that will succeed are those that pair decisiveness with precision, stripping away the distortions

3: When Data Becomes Strategy

The war room glows with walls of screens — markets shifting in one corner, customer feedback scrolling in another, predictive models updating by the second. Executives no longer argue about whether data is available; they wrestle with what it means, how to act, and how to align. Instinct takes a back seat. The real debate is which insight carries the most weight for the next move. Strategy is no longer drawn from hindsight or operational support. It emerges when data itself becomes the foundation of direction.

For decades, strategy was shaped by broad strokes. Leaders studied industry reports, hired consultants, and relied on gut to fill the gaps. The numbers were often old by the time they reached the table. That world is gone. Today, companies generate a torrent of live information from sensors, apps, transactions, and customer interactions. The problem is no longer scarcity — it is abundance. The winners are those who turn abundance into clarity, and clarity into action.

Data is no longer back-office support. It is front-line strategy. Consider Walmart. For years it was seen as a brick-and-mortar giant. Yet Walmart now analyzes petabytes of transaction data daily, feeding AI models that predict demand down to the store level. When hurricanes approach, the system doesn't just stock flashlights — it knows strawberry Pop-Tarts will sell out. That insight isn't trivia. It translates into revenue, loyalty, and market share. Strategy is embedded in the dataset.

Amazon is another example. Its recommendation engine is often described as marketing, but it is strategy in action. By tracking customer clicks, searches, and purchases, the system influences what people buy before they realize they want it. That's not support. That is the business model itself. Every product line, from cloud computing to groceries, builds on the same foundation: turning raw signals into decisions about what to sell, how to price, and where to expand.

These examples highlight the shift executives must face. Strategy is no longer a static document updated once a year. It is a living process shaped by continuous streams of input. For companies that fail to grasp this, the risk is falling behind while rivals make faster, smarter moves.

Executives often ask how to make this shift without drowning in dashboards. The answer is not more reports. It is a different way of thinking about data as a strategic asset. That begins with a few key steps:

- Treat data as infrastructure, not exhaust. Build platforms that collect, clean, and connect across silos.

- Move from descriptive reporting to predictive and prescriptive analytics. Don't just know what happened — see what is likely to happen next.

- Put AI at the center of decision-making, not at the edges. Let it propose actions, not just metrics.

- Tie executive incentives to data-driven outcomes, so leaders don't fall back on instinct when evidence points elsewhere.

Each of these steps requires cultural and operational change. But each is necessary if data is to become the heart of strategy.

One area where this is already visible is healthcare. Hospitals generate massive volumes of data but historically struggled to use it. Today, AI-driven platforms analyze patient records, treatment outcomes, and supply needs. They guide staffing decisions, predict surges in demand, and identify early warning signs for disease. For leaders in healthcare, this is no longer an IT function — it is strategic planning. Where to open new clinics, how to allocate funding, which treatments to prioritize — these are strategy questions answered through data.

Finance shows another angle. Asset managers once relied heavily on quarterly reports and analyst calls. Now, firms track satellite images of retail parking lots to predict sales, scrape social media to measure sentiment, and use machine learning to spot correlations no human could find. When BlackRock introduced Aladdin, its risk and investment system, it changed the way decisions were made across the firm. Aladdin processes millions of simulations daily, shaping portfolio strategy in ways no traditional committee could match.

Executives who want to harness this shift must rethink how their organizations are structured. Too many still treat data as the domain of IT. That model breaks down when data is the source of advantage. Strategy teams must be built with data science at the core, not as a support function. Every major decision—from product launches to acquisitions—should begin with models that test outcomes under varied scenarios. Human judgment still matters, but judgment should be informed by simulation, not by instinct alone.

There is also a cultural challenge. Leaders often cling to intuition because it feels personal. They want to be seen as visionaries. Yet vision without evidence is fragile. A forward-looking executive should frame vision as the ability to see what data already shows, then act on it before others do. That's not giving up leadership. It is redefining leadership for an era when information is too vast for any one mind to process.

Some executives worry this makes strategy too rigid. They fear becoming slaves to algorithms. That risk is real if data is treated as gospel. But the real opportunity lies in treating data as a conversation partner. AI proposes. Leaders interpret, adapt, and communicate. This partnership allows companies to move faster without losing context. The discipline comes from letting data shape the agenda, not from abandoning human perspective.

Implementation requires discipline. A practical roadmap for executives could include:

- Conduct a full audit of corporate data assets to identify gaps and overlaps.

- Invest in platforms that unify data across operations, finance, HR, and customer experience.

- Establish a decision hub where AI-driven insights are reviewed alongside human judgment before action.

- Train executives in data literacy so they can question, challenge, and explain outputs to stakeholders.

- Build feedback loops that measure the success of data-driven strategies and refine models over time.

These steps turn abstract talk into operational reality. They also create accountability. If an AI-driven pricing recommendation lifts revenue by 5 percent, leaders can point to evidence, not anecdote. If a supply chain algorithm reduces costs by 10 percent, the board can track the result. Accountability strengthens when outcomes are tied to data rather than stories.

The shift to data-driven strategy is not uniform. Some industries embrace it faster than others. Retail, logistics, and finance are already far ahead. Manufacturing is catching up through predictive maintenance and sensor data. Education and government lag, but pressure is growing. Wherever competition is sharpest, data-driven strategy rises fastest. The signal is clear: companies that wait will be outpaced by those who act.

Scholars often debate whether strategy is art or science. For decades, the argument leaned toward art — vision, charisma, and storytelling. But the evidence is shifting. Science is taking center stage. Data allows companies to test hundreds of strategic options before committing. It exposes hidden risks and hidden opportunities. It transforms leadership from an act of personality into a process of disciplined choice.

For the executive reader, the challenge is personal. Ask yourself: how much of your current strategy is rooted in data, and how much rests on instinct? How often do you rely on last year's playbook rather than models that show what's coming next? How often do politics in your organization warp the signal? These are uncomfortable questions, but they are necessary if you want to lead in an era where advantage comes from clarity.

Data is no longer a byproduct. It is the raw material of strategy. Those who harness it will set direction with confidence. Those who ignore it will find themselves chasing competitors who act faster and with greater precision. The shift is already underway, and it favors the leaders who see that strategy itself has changed.

"The end of executive elites is not the end of leadership. It is the start of shared authority."
— *Dave Karpinsky*

Part II: Why the CEO is Next

4: From Assistants to Executives

The first time many executives encountered AI, it wasn't in a boardroom. It was on their phones. Virtual assistants scheduled meetings, answered simple queries, and reminded them of tasks. These tools were helpful but not transformative. They felt like conveniences, not leadership. Yet those early assistants marked the beginning of a shift. Over time, AI stopped being a background utility and began moving closer to the center of corporate authority. The journey from scheduling emails to shaping board-level strategy has been steady, incremental, and often underestimated.

In the early years, chatbots were the most visible face of AI in business. Banks deployed them for customer service. Airlines used them for flight changes. Retailers installed them to answer basic product questions. Most executives saw them as cost-saving tools, nothing more. But even these simple bots offered lessons. They showed how AI could interact with people, learn from feedback, and improve with scale. More importantly, they acclimated employees and customers to the idea of speaking with a system rather than a person. That cultural adjustment paved the way for deeper adoption.

The next phase came through analytics. Companies realized their data warehouses were overflowing. Spreadsheets and dashboards couldn't keep up. AI entered to detect patterns, flag risks, and forecast demand. For example, airlines like Delta and United began using AI to optimize pricing in real time. Instead of quarterly pricing committees, algorithms adjusted fares by the hour. This wasn't just operational support. It was an executive function — allocating revenue streams more precisely than humans could. The label was analytics, but the reality was authority.

Over time, AI shifted from reporting what had happened to predicting what was likely to happen. Predictive analytics gave leaders foresight they never had before. Retailers started stocking products based not on last year's sales but on signals from search data, weather, and local trends. Logistics firms restructured shipping schedules weeks ahead based on predictive models. In each case, decisions once made by executives were quietly automated. Few noticed at first. It felt incremental. Yet step by step, the balance of power moved.

By the mid-2010s, machine learning models were being trained on vast datasets that dwarfed anything humans could analyze. Credit risk, fraud detection, and insurance underwriting all shifted into AI systems. At JPMorgan, contract review once required thousands of lawyer-hours. AI tools now scan agreements in seconds, flagging risks and inconsistencies. That's not administrative support. That's executive judgment, delegated to code. For leaders, the message was clear. AI wasn't just assisting. It was deciding.

A turning point came with the rise of board-level analytics. Boards began demanding independent data streams, not just curated slides from management. AI-driven systems provided that. Directors could run scenarios, test assumptions, and validate claims without relying on the CEO alone. This reduced the information asymmetry that had long favored executives. It shifted the dynamic. AI became not just a servant of management but a trusted advisor to the board. That was a new level of authority.

Executives watching this shift should ask themselves how incremental adoption turns into structural change. The lesson is simple: AI creeps upward by delivering results where humans are slow, inconsistent, or biased. The more accurate it becomes, the more authority it earns. Over time, the scope expands. What begins as a chatbot ends as a voice at the table.

The pattern is consistent across industries. In healthcare, AI began by scanning images to flag anomalies for radiologists. Now it helps hospital administrators allocate staff, predict patient surges, and optimize budgets. In supply chains, AI started with route planning. Now it determines where warehouses should be built and how contracts should be negotiated. In retail, AI moved from recommending products online to deciding how much inventory to allocate by region. These are executive calls — resource allocation, risk management, and growth planning — no longer left to human leaders alone.

Executives must learn to guide this transition rather than resist it. That requires intentional steps. Begin by identifying areas where AI has already proven reliable at operational levels. Look for patterns of accuracy and repeatability. Then expand its authority upward. For example:

- If AI has proven accurate in predicting customer demand, allow it to inform regional expansion strategies.

- If AI consistently reduces fraud detection errors, let it influence policy decisions on credit approval frameworks.

- If AI pricing models outperform committees, tie executive bonuses to the adoption of algorithm-driven recommendations.

Each step builds trust. Each expansion shows that authority once thought untouchable can be shared or transferred.

Another practical activity is building mixed decision forums. Too often, AI outputs are presented as reports at the end of meetings. Instead, make them the opening. Begin strategic discussions with AI recommendations, then allow human debate to follow. This flips the sequence. It signals that AI is not just commentary — it is agenda-setting. Over time, that cultural shift makes AI a peer, not a servant.

Executives should also invest in explainability. Authority grows when systems are trusted. If leaders can't explain AI outputs, trust collapses. This means building models that produce not only answers but reasoning chains. For instance, if an AI recommends divesting from a region, it should show the demand forecasts, regulatory risks, and capital costs that shaped the call. Explainability is the bridge that takes AI from assistant to executive. Without it, adoption will stall.

Boards face their own tasks. They must expand governance frameworks to include algorithmic oversight. This means defining clear lines of accountability. If AI recommends an acquisition, who signs off? If AI forecasts risk that management ignores, how is that documented? Boards should demand transparency into both AI performance metrics and management's response. Over time, this creates a dual accountability system: executives answer to the board, and AI outputs shape the board's oversight.

The scholarly debate here is fascinating. Some argue incremental AI adoption is dangerous because it creates "automation creep" — small delegations that add up to loss of control. Others argue the opposite: that incremental adoption is the safest path, because it allows culture and governance to adapt step by step. Both perspectives matter. What cannot be ignored is that the process is underway, and resisting it only delays the inevitable.

For executives who want to act, a phased implementation roadmap helps. Consider the following:

- Phase 1: Operational delegation – Give AI authority over repeatable, high-volume tasks like scheduling, routing, or transaction monitoring.

- Phase 2: Tactical delegation – Expand to decisions that shape revenue and cost structures, such as pricing, staffing, or forecasting.

- Phase 3: Strategic delegation – Allow AI to propose and defend moves like market entry, capital allocation, or portfolio diversification.

- Phase 4: Executive integration – Position AI outputs alongside human recommendations in board and leadership meetings.

Each phase requires clear benchmarks. Performance must be tracked and validated. Human override should remain possible, but not automatic. Trust grows when leaders see that AI-driven decisions consistently outperform human-only calls.

The cultural aspect cannot be overstated. Employees must see AI as credible. That requires communication. When Deutsche Bahn expanded AI-driven scheduling in its rail networks, it paired every rollout with town halls and training sessions for staff. Leaders explained not just what the system did, but why. They showed workers how outputs reduced errors and improved safety. Adoption improved because people understood the purpose. For executives, the lesson is clear. Technology adoption is not only technical. It is cultural.

The movement from assistants to executives is still in motion, but the trajectory is unmistakable. What begins as help with small tasks steadily scales into authority over major decisions. Each success makes resistance harder. Each win builds credibility. Leaders who treat AI as a permanent assistant will find themselves bypassed by competitors who treat it as an emerging executive. The choice is not whether AI climbs the hierarchy. The choice is how prepared you are when it does.

5: Unbiased Decisions, Unlimited Scale

A leadership meeting drags on late into the night. Reports pile high, market updates flood in, and the weight of expectation hangs over the room. Executives are tired. Some arrive with their minds already made up. Others hesitate, protecting teams or reputations. Everyone knows the choice will ripple through the company. Yet the outcome isn't shaped by data alone. It bends under human limits — fatigue, ego, and politics. Now place in that same room a participant immune to those pressures. A system that can process every datapoint, run thousands of simulations, and deliver a recommendation untouched by bias. That is the promise of AI at scale.

The most powerful argument for AI in executive roles is its ability to make decisions without human distortions. Leaders like to think they are objective. Decades of research shows otherwise. Cognitive biases, from anchoring to overconfidence, shape judgment in ways people rarely notice. Studies by behavioral economists reveal how consistently executives fall into these traps. The risk is not lack of intelligence. It is the unavoidable limits of human thinking. AI approaches problems differently. It doesn't grow tired after long debates. It doesn't cling to legacy beliefs or personal pride. It doesn't soften facts to avoid conflict. When designed well, it brings clarity that humans alone cannot match.

Take the challenge of processing scale. Global corporations operate across dozens of markets, thousands of suppliers, and millions of customers. Human executives can review a few reports at once. AI can review them all. For example, Google's ad pricing is not set by committees in weekly meetings. It is managed by algorithms that analyze trillions of signals each day, adjusting in milliseconds. Those decisions shape billions in revenue daily. No human could hope to operate at that speed or breadth. The principle extends to other sectors. Airlines use AI systems that weigh weather, fuel costs, booking patterns, and geopolitical risks in real time. Manufacturing firms now run predictive systems that monitor equipment across hundreds of plants at once, alerting managers before breakdowns occur. Scale is no longer a barrier when decisions are data-driven.

Executives should reflect on what scale means in their own organizations. In many firms, decisions about capital allocation, pricing, or expansion are constrained by the capacity of teams to review inputs. Imagine shifting that burden to systems that never tire and never forget. The human role shifts from collecting and summarizing to questioning and guiding. The bottleneck is removed. Decisions move from weeks to minutes.

Bias is the other great weakness of human leadership. Even seasoned executives are swayed by relationships, reputations, and their own history. Consider the sunk cost fallacy — the tendency to keep funding a project because of past investment, even when evidence says stop. AI does not care about sunk costs. It reviews the current data, recalculates, and proposes the best path forward. That discipline can save billions. In energy markets, algorithmic trading platforms often recommend pulling out of positions that humans cling to out of pride or hope. The systems are cold, but effective. They are not distracted by emotions that have undone many great leaders.

Ego is another distortion. Human leaders often fear looking indecisive. They double down when proven wrong. AI has no ego to protect. It updates its recommendation as soon as new data arrives. During the pandemic, supply chains shifted daily. Human-led committees struggled to adjust in time. AI systems at firms like DHL and UPS re-routed shipments and recalibrated logistics hour by hour. Those decisions weren't about pride. They were about math. And they worked.

Executives who want to apply these lessons should take deliberate steps to integrate AI into their decision frameworks. Start by identifying areas where bias and fatigue are most costly. Common examples include:

- Capital allocation across regions or product lines

- Pricing decisions in volatile markets

- Talent management, from hiring to promotion

- Supply chain routing under disruption

- Risk assessment for compliance and regulation

Next, embed AI systems to generate independent recommendations. Don't just ask for support metrics. Ask for decisions. Present these alongside human options in executive meetings. This creates a counterweight to politics. If a leader advocates for a strategy, the AI perspective sits next to it. The debate must address evidence, not just narrative.

Boards can reinforce this by demanding to see both human and AI-generated recommendations. This creates transparency. If management ignores the AI option, the board can ask why. That pressure alone reduces the influence of bias. It forces executives to confront their limits and explain their choices.

One of the greatest strengths of AI is its consistency. Human decision-making varies by mood, time of day, and personal interest. Studies show judges give harsher sentences before lunch than after. Sales managers are often more generous with discounts late in the quarter. CEOs under stress take fewer risks, even when risk is warranted. AI does not swing with mood. It delivers the same analysis at midnight as at noon. It applies the same thresholds across regions, departments, and time zones. That consistency builds fairness, which strengthens trust across an organization.

Consider the hiring process. Human interviews are notoriously inconsistent. One candidate may be judged more harshly because the interviewer is distracted. Another may benefit from unconscious bias. AI-driven assessments, when designed properly, apply the same standards to every applicant. This doesn't eliminate bias entirely — data quality matters — but it creates a more even field than human-only processes. Companies that adopt these systems report more diverse and better-performing hires. For executives, that lesson extends beyond HR. Consistency in decisions drives consistency in outcomes.

The unlimited scale of AI also allows organizations to simulate outcomes that humans cannot. Scenario planning has always been part of executive life. But human teams can only model a handful of possibilities at once. AI can run millions. When Shell adopted AI-driven modeling for energy markets, it was able to simulate global demand scenarios across decades, testing the impact of carbon taxes, technology shifts, and regional instability. Human strategists set direction, but the system gave them foresight they never had before. That foresight translated into strategic advantage.

Executives can take practical steps to replicate this approach. Build AI-driven scenario models for key strategic decisions. For instance:

- When considering expansion into a new country, run models on regulatory shifts, currency swings, and competitor reactions.

- When evaluating an acquisition, test scenarios for integration speed, customer retention, and cultural alignment.

- When setting long-term pricing strategies, simulate impacts of inflation, wage pressure, and raw material volatility.

These models won't replace judgment, but they will expand the horizon of possibilities. Leaders who rely only on human teams are limited by bandwidth. Leaders who rely on AI gain a broader field of vision.

Scholars debate whether reliance on AI creates passivity in leaders. Will executives stop thinking critically if the system does the analysis? This is a valid concern. The counterpoint is that AI frees leaders from cognitive load so they can focus on interpretation, communication, and ethical framing. The machine can run the math. The human must decide how to explain it, how to balance competing priorities, and how to connect the decision to culture. Far from erasing leadership, this shift elevates it.

Executives must also prepare their organizations for the cultural impact of unbiased decision-making. People are accustomed to persuading leaders, lobbying for outcomes, and building alliances. When AI enters the room, those dynamics shift. Some may feel threatened when politics lose power. Others may feel relieved that decisions rest more on data than influence. Leaders should anticipate both reactions. The key is communication. Explain how AI contributes, where it has authority, and how employees should interact with its outputs. Transparency reduces fear and builds trust.

One practical method is to create "AI decision logs." These record the recommendation made by AI, the final human decision, and the outcome. Over time, the logs show patterns. If AI consistently outperforms human choices, it strengthens the case for expanding its authority. If AI fails, the logs reveal where improvements are needed. This approach mirrors how financial audits build trust. It creates accountability without emotion.

Another method is phased delegation. Don't hand over authority all at once. Begin with narrow decisions where bias and fatigue have high costs. Expand as confidence grows. For example, allow AI to set inventory levels before granting it authority over capital projects. Allow AI to adjust regional pricing before expanding to corporate mergers. Each success builds confidence in the next step.

AI also enables organizations to make decisions faster without sacrificing accuracy. Human committees often delay action to gather more information or build consensus. AI can provide a recommendation instantly. This speed can be decisive in markets where timing matters. In e-commerce, milliseconds of delay can mean millions in lost sales. In finance, trading decisions must be made in microseconds. In healthcare, treatment recommendations must be delivered immediately. Speed is not a luxury. It is survival.

Executives should ask themselves how speed could change their own industries. Could faster pricing decisions capture more margin? Could quicker supply chain adjustments reduce shortages? Could real-time customer feedback shape product design before competitors respond? AI provides that speed without the compromises that humans often make under pressure.

Finally, the scale and lack of bias in AI decisions opens the door to new levels of personalization. Humans cannot tailor decisions to millions of customers at once. AI can. Netflix doesn't offer one recommendation for all users. It offers millions of unique recommendations, updated constantly. That level of personalization would be impossible for any human team. For executives, the lesson is clear. The same principle can apply to banking, healthcare, education, and beyond. Strategy is no longer about averages. It is about individualized experiences at scale.

The story of unbiased decisions and unlimited scale is not about replacing humans. It is about addressing the flaws of human leadership that have long been accepted as unavoidable. AI demonstrates that fatigue, ego, and politics are not inevitable. They are correctable. The future belongs to leaders who embrace that correction and build systems that turn clarity into advantage.

6: The Cost Advantage: A CEO That Doesn't Get Paid

Picture a compensation committee reviewing executive pay. The numbers are staggering: multimillion-dollar salaries, annual bonuses, golden parachutes, equity awards that stretch into nine figures. Every year, headlines spark debate about whether CEOs are worth the cost. Boards argue that the market demands these packages to attract and retain talent. Shareholders grumble about dilution and overhead. Employees watch from below, comparing their wages to those of leaders who earn in a single day what they make in a year. The scale of executive pay has become both a financial and cultural flashpoint. Now imagine a different conversation. What if the CEO didn't draw a salary at all? What if the role was filled not by a person but by a system — an AI leader immune to compensation negotiations, bonuses, or perks? The potential cost advantage is enormous, not just in pay savings but in how resources could be reallocated across the enterprise.

Executive pay has climbed sharply in recent decades. In the United States, the ratio of CEO pay to average worker pay rose from 20-to-1 in the 1960s to more than 350-to-1 today. Many global corporations pay their top executives tens of millions annually, with additional equity incentives worth far more. These numbers aren't small line items. They shape investor relations, influence stock buybacks, and frame internal morale. For boards under pressure, the idea of eliminating that expense is hard to ignore. AI-driven leadership offers that possibility.

The financial savings extend beyond headline salaries. Corporate overhead tied to executives includes private jets, security teams, exclusive retreats, deferred compensation, and severance packages that often run into tens of millions. The 2017 exit of GE's CEO, Jeff Immelt, cost the company nearly $200 million when severance, pensions, and benefits were included. These packages aren't unusual. They reflect a market where boards compete for talent with ever-larger rewards. AI doesn't ask for severance. It doesn't demand perks. It delivers decisions without drawing down shareholder equity.

The counterargument is clear. Supporters of high executive pay argue that leadership quality justifies the cost. A great CEO, they claim, adds billions in value through vision, strategy, and execution. Paying $20 million a year is small compared to market capitalization gains. Yet evidence is mixed. Studies from Harvard and Stanford show that CEO pay often has little correlation with long-term performance. Short-term incentives push leaders toward stock buybacks, aggressive cost-cutting, or risky acquisitions that lift share price in the moment but damage companies over time. The link between pay and performance is not as strong as defenders suggest. AI sidesteps this by aligning decisions directly with performance outcomes rather than personal compensation targets.

Consider how incentives distort decisions. A CEO with large stock options may favor buybacks to lift share prices, even when investing in innovation would create greater long-term value. An AI system, programmed with clear corporate objectives, would weigh trade-offs without self-interest. It would prioritize based on data, not on how the outcome affects its net worth. That difference is not just a moral advantage. It is a financial one. Companies lose billions each year to misaligned incentives. Removing those distortions can be as valuable as eliminating pay packages themselves.

Executives reading this should think beyond cost savings to structural reallocation. If a corporation redirected $20 million in annual CEO pay, what could be achieved? Some options are obvious:

- Invest in employee training programs to boost retention and productivity.

- Fund R&D projects that were shelved due to budget constraints.

- Expand sustainability initiatives that reduce regulatory risk and improve reputation.

- Increase dividends or share buybacks in ways aligned with long-term value.

- Strengthen cybersecurity and IT infrastructure, areas often underfunded until after a breach.

Each dollar no longer spent on executive pay becomes capital that can strengthen the enterprise. The return on that reinvestment could exceed the supposed "value add" of a traditional CEO.

Another overlooked advantage lies in predictability. Human executives negotiate pay annually, benchmarking against peers and often threatening to leave if packages aren't competitive. This creates cycles of escalating compensation across industries. AI doesn't negotiate. It doesn't require contracts or retention bonuses. Its cost is stable: licensing, maintenance, and development. That predictability simplifies financial planning. It reduces volatility in overhead. For boards and investors, stability itself is value.

The cost advantage extends further into organizational structure. Many companies build entire HR, legal, and consulting functions around executive compensation. Compensation committees hire advisors to benchmark pay. Consultants produce reports justifying packages. Lawyers draft contracts. Each of these processes costs millions. Replace the CEO with an AI system, and the ecosystem of compensation management shrinks. The savings multiply beyond the top line.

Skeptics point out that developing and maintaining AI leadership systems also costs money. They are correct. But the scale is different. Even advanced AI platforms with custom modeling, governance layers, and cybersecurity protections cost a fraction of executive compensation over time. A robust system might cost tens of millions to build and a few million annually to maintain. Spread across years, that remains below the recurring pay of human executives. Unlike human pay, system costs don't escalate with market cycles. They decrease as technology improves.

Executives can implement this advantage by starting small. They should begin by calculating total executive compensation — not just salaries, but bonuses, equity grants, severance risk, perks, and overhead. This "true cost" often shocks boards once fully tallied. Then compare that to projected AI system costs, including hardware, software, and staff to monitor performance. Present the gap as an opportunity for reinvestment. By framing the conversation around capital allocation rather than personality, boards can take a rational step toward adoption.

Boards should also run scenarios on reinvestment. If $30 million in annual compensation were redirected, what measurable outcomes could be expected? Simulations can show whether reinvestment in training, R&D, or dividends produces higher returns than current compensation practices. Framing AI adoption as an investment strategy, not just a technology experiment, aligns with shareholder priorities.

The scholarly debate here is intense. Critics argue that removing pay incentives from leadership risks weakening performance. Without personal reward, they say, motivation drops. Yet AI doesn't operate on motivation. It operates on goals. Performance is tied directly to design, not desire. The challenge is ensuring that design aligns with corporate strategy and ethics. This is where boards and stakeholders must step in. By defining objectives clearly, they shape the incentives of the system itself. Unlike with human leaders, those incentives won't drift over time.

Another consideration is fairness. Employee morale suffers when workers see executives earning hundreds of times more than they do. Replacing paid CEOs with AI could reduce tension across organizations. It signals a shift toward equality of purpose. When compensation gaps narrow, engagement often rises. Employees feel that value is distributed more evenly. That cultural dividend may be as significant as financial savings.

Executives can prepare their companies for this cultural impact by pairing AI adoption with communication campaigns. Show employees how savings are reinvested. Make clear that funds once reserved for executive perks now support employee growth, innovation, or benefits. Transparency builds trust. Without it, skepticism grows. Workers may fear that AI adoption is just another cost-cutting move. With transparency, they see the purpose and the benefit.

AI leadership also avoids the reputational risk tied to excessive pay. Every year, headlines highlight CEOs receiving record-breaking packages while employees face layoffs or wage freezes. Public backlash harms brand reputation and shareholder relations. AI removes that risk. No journalist will report outrage about a system that doesn't draw a salary. That reputational shield is another form of value, protecting companies from distraction and criticism.

Boards should be proactive here. Public companies already face pressure from activist investors and regulators over executive pay. By adopting AI leadership, they can position themselves as forward-looking and fiscally responsible. That positioning can become part of investor relations, highlighting how the company allocates capital with discipline. It sets a new standard for governance, moving away from personality-driven compensation toward performance-driven systems.

The cost advantage also strengthens during crises. When markets crash or demand falls, boards often face backlash for paying executives large bonuses while cutting staff. AI leaders avoid that dilemma. They don't require bonuses to stay motivated. They don't demand retention packages to weather downturns. Their focus is on stability and survival, not personal gain. That alignment with corporate needs creates resilience during shocks.

Implementation requires careful planning. Executives and boards should:

- Conduct a full compensation audit to capture the true cost of executive leadership.

- Benchmark AI system costs over five- and ten-year horizons.

- Build reinvestment models to show potential shareholder returns.

- Establish governance frameworks for AI oversight in place of compensation committees.

- Communicate savings and reinvestment to employees, investors, and regulators.

This process reframes leadership costs as a strategic choice rather than a market obligation. Once boards see the scale of savings, momentum for change grows.

The transition won't be simple. Leadership pay is deeply tied to identity, prestige, and status. Many executives resist reform because they benefit from the system. But the financial argument for AI is too strong to ignore. When corporations can cut tens of millions in overhead while improving decision accuracy, the case builds itself. Investors will press for adoption. Boards will look for savings. Employees will welcome fairness. The market will shift toward systems that don't need paychecks.

"A company guided by AI does not erase human
purpose—it demands it more than ever."

— *Dave Karpinsky*

Part III: The Shape of AI-Driven Corporations

7: AI as the New Corporate Brain

Think of a corporation as a body, with leadership serving as its brain. For decades, that brain has struggled to keep up. Finance delivered one version of the numbers, operations another, HR another, and market intelligence added its own view. Leaders tried to integrate it all, but the connections were weak. Silos slowed communication. Signals were filtered through politics, fatigue, and selective reporting. Today, a new kind of brain is emerging—one that can process every signal at once. AI is stepping into the role of the central nervous system for the modern enterprise.

Most companies today still run on fragmented intelligence. Finance runs forecasts in spreadsheets. Operations manage supply chains with separate platforms. HR tracks engagement in surveys and payroll systems. Marketing interprets consumer trends from social feeds and loyalty programs. These silos force executives to act as integrators. They sit in meetings, listen to conflicting interpretations, and try to balance them into strategy. That process is slow, reactive, and error-prone. AI promises to change this by becoming the hub where all corporate data converges.

Think about the flow of signals. Finance wants to know how costs will shift if raw materials spike. Operations wants to know whether suppliers can keep pace. HR wants to know if staff are stretched too thin. Marketing wants to know how consumer behavior will shift under higher prices. Each of these questions is linked, but today they are handled in isolation. AI integration allows these signals to be processed together. The system can weigh trade-offs instantly: raising prices may increase margin but could strain operations and require more staffing. Leaders no longer need to juggle siloed reports. They receive a unified view.

This vision isn't theory. Some companies already move in this direction. Siemens runs AI platforms that integrate data across plants, suppliers, and markets, providing executives with end-to-end visibility. HSBC integrates compliance, risk, and finance data into AI hubs that flag risks across business units in real time. Tesla uses integrated AI systems that connect production, software updates, and customer data, creating feedback loops from driver behavior to design changes. These examples show how integration reshapes not just efficiency but decision-making authority itself.

Executives should ask: what would change if every part of the enterprise were connected into a single decision brain? The answer is speed and clarity. Instead of waiting weeks for quarterly reviews, leaders could see live trade-offs across the entire organization. Decisions wouldn't just be faster. They would be more aligned, since the system sees connections humans miss.

The shift begins with data integration. AI cannot act as a brain if inputs remain fragmented. That means companies must:

- Audit existing systems to identify where finance, HR, operations, and marketing collect data separately.

- Build data pipelines that connect these systems into a central platform.

- Standardize data definitions so the AI interprets payroll hours the same way across regions, or supply delays the same way across vendors.

- Invest in real-time feeds, replacing quarterly reporting with continuous updates.

Once integrated, AI becomes more than a tool. It becomes a decision hub. For example, consider workforce planning. HR systems may show a shortage of skilled engineers in one region. Finance systems may show the cost of hiring rising sharply. Operations may show increased demand for production in that same region. Traditionally, these insights arrive at separate times, forcing executives to make delayed choices. An AI brain can process them simultaneously, recommend hiring in another region, and adjust supply chain forecasts accordingly. What once took months can occur in minutes.

The scholarly debate here focuses on control. Critics argue that handing integration to AI risks making decisions too mechanistic. Human leaders may feel sidelined. Yet defenders counter that AI doesn't remove human purpose. It removes friction. The brain does not decide goals. It processes signals and executes responses. In this framing, AI empowers leaders by giving them clarity and freeing them to focus on meaning, ethics, and long-term direction.

One of the most transformative advantages of an AI corporate brain is the elimination of hidden costs created by silos. In many firms, projects are approved by finance without a clear view of HR capacity. Operations expands without seeing the impact on working capital. Marketing launches campaigns without seeing the effect on supply chains. These disconnects waste billions annually. AI integration surfaces these conflicts in real time. It prevents initiatives from colliding. It aligns the body by aligning the brain.

Executives who want to implement this must focus on practical steps. Start by building "integration dashboards" powered by AI. These should not be traditional dashboards with static metrics. They should be live systems that update constantly and highlight cross-functional impacts. For example:

- A pricing change in finance should immediately show projected effects on customer churn from marketing data.

- A new plant opening in operations should instantly show staffing requirements in HR data.

- A new regulatory rule in compliance should update projected capital impacts in finance.

These integrations move beyond reporting into decision orchestration. They position AI not as a department tool but as a corporate brain.

The scale of benefit grows when external data is included. Markets don't wait for quarterly meetings. Competitors launch products, regulations shift, and supply chains face shocks daily. By feeding external data into the corporate brain, companies gain foresight. An AI hub can combine internal sales with global market data, predicting how currency changes will affect demand. It can combine supplier delays with geopolitical signals, predicting risks before executives even meet. This expands the brain from inward focus to outward vision.

Executives must also address governance. A corporate brain with access to finance, operations, HR, and market data must be monitored with care. Governance frameworks should ensure data privacy, ethical use, and accountability. Boards should create AI oversight committees, much like audit committees today. These committees should review system decisions, track accuracy, and monitor risks of unintended outcomes. Governance ensures that integration enhances trust rather than erodes it.

Another layer of opportunity lies in personalization. A corporate brain can provide tailored insights for each leader. The CFO may see capital impacts first, the COO may see operational risks, and the CHRO may see workforce needs. All draw from the same hub, but each view is personalized. This reduces misalignment across leadership teams. Everyone works from one source of truth, but with insights tailored to their function. The result is harmony in decision-making.

Executives should recognize that AI as the corporate brain also changes culture. Decisions shift from persuasion to evidence. In traditional meetings, executives argue from different reports, lobbying for their position. With AI integration, the system provides a common baseline. Debate shifts to interpretation and strategy, not to data quality. That cultural change reduces politics and increases focus. It reshapes how leaders spend their time.

Practical implementation requires leaders to rethink metrics. Traditional KPIs often reflect silos. Finance tracks margin, HR tracks turnover, operations track throughput. An AI brain can create integrated KPIs. For example:

- "Customer profit per employee hour" combines finance, HR, and marketing data.

- "Capital efficiency per supply unit" combines finance and operations.

- "Time-to-market per dollar invested" combines R&D, finance, and sales.

These integrated metrics give a truer picture of performance. They prevent departments from optimizing locally at the expense of the enterprise. They align the brain toward common goals.

The debate among scholars is not whether integration is possible, but whether it risks overcentralization. Some argue that too much reliance on one system could reduce adaptability. Others argue the opposite: that integration creates agility because responses are coordinated. For executives, the lesson is balance. Don't centralize for the sake of control. Centralize to improve clarity, while keeping room for human flexibility.

Another consideration is scalability. AI brains can grow as the company grows. When entering a new market, data streams from that market can be plugged into the hub, instantly connected to finance and operations. When acquiring another company, integration can occur faster by connecting datasets to the corporate brain rather than merging systems manually. This scalability reduces the cost and complexity of growth.

Executives can apply these insights through structured rollouts:

- Phase 1: Connect internal silos (finance, HR, operations, marketing).

- Phase 2: Add external data (competitors, regulators, suppliers).

- Phase 3: Shift from descriptive to prescriptive outputs — recommendations, not just reports.

- Phase 4: Build feedback loops that compare outcomes against AI recommendations, refining accuracy.

Each phase should be measured with clear ROI metrics. Savings from reduced waste, faster decisions, and avoided conflicts can be tracked. These outcomes build confidence in the system.

The metaphor of the corporate brain is powerful because it reframes leadership. The body does not argue with the brain about whether signals should connect. It trusts the brain to integrate inputs and coordinate responses. For corporations, AI offers the same possibility. Integration replaces friction. Clarity replaces confusion. Leaders are freed to focus not on stitching signals together but on defining direction.

8: Redefining the Role of Humans

In the boardroom, decisions now move at machine speed. Forecasts refresh in real time, risks are flagged instantly, and resources shift without drawn-out debate. An AI system presents strategy with precision, pulling insight from every corner of the company and the market. Executives watch as clarity unfolds across the screen—flawless, logical, efficient. Then the pause arrives. How will this decision be explained to employees, investors, and regulators? How will the message be told in a way that builds trust? That is where humans step in.

The rise of AI in leadership doesn't remove the need for people. It changes where people matter most. The role of human executives shifts from being decision-makers burdened by bias and fatigue to being interpreters, communicators, and connectors. This is not about competing with AI but complementing it. Machines can process more data than any human mind. But machines cannot replace empathy, trust, and the ability to inspire.

One of the most enduring human roles is storytelling. Strategy without story is cold. It may be correct, but it doesn't connect. Employees don't rally around algorithms; they rally around meaning. Investors don't commit capital just to numbers; they commit because they believe in a vision. Regulators don't respond to data alone; they respond to context. Storytelling weaves facts into narrative.

It helps people understand why choices matter and how they fit into a larger purpose.

Think of how Apple under Steve Jobs framed products. The iPod was not sold as a technical device with storage capacity. It was presented as "1,000 songs in your pocket." That framing shaped culture, not just sales. An AI system could analyze markets and predict demand. But it could not create that story. It could not spark imagination in the same way. Humans retain that role. Executives must learn to craft stories that make sense of AI-driven strategies, giving them meaning employees and customers can grasp.

Another uniquely human role is empathy. AI can detect sentiment in surveys or tone in text. But it doesn't feel the weight of disappointment, pride, or fear. Leaders must. When a factory closes, the numbers may justify it. The AI may show efficiency gains and higher margins. But people see jobs lost, communities disrupted, and futures uncertain. Only human leaders can stand before those communities, acknowledge the pain, and connect with dignity. Empathy is not weakness in leadership. It is strength. It builds trust, which makes adoption of AI-driven strategies possible.

Negotiation is another domain where humans remain vital. Data can suggest optimal outcomes, but negotiation is rarely about pure optimization. It involves pride, face-saving, and unspoken signals. A merger agreement may hinge not on valuation but on how founders feel their legacy is treated. A labor contract may hinge not on pay but on dignity. An AI might propose the "perfect" settlement on paper. But without human negotiators who read the room, sense the dynamics, and build compromise, deals can collapse. Negotiation remains one of the strongest human skills in executive life.

Vision-setting also belongs to humans. AI excels at projecting from the past and modeling the present. But vision is about imagining futures that don't exist yet. It is about framing a company not only around what is likely but around what is possible. Vision requires daring, creativity, and sometimes irrational belief. History shows that great companies are often built on leaps of imagination that data alone would not have justified at the time. AI can support vision, but it cannot create it. Leaders must hold onto this role as their most critical contribution.

Executives should reflect on how to implement these roles practically. Start by redefining leadership development programs. Instead of teaching managers to master spreadsheets or memorize financial models, train them in communication, empathy, and negotiation. Build curricula around storytelling for strategy, listening skills for leadership, and framing vision for growth. These are the skills that will complement AI, not compete with it.

Another action is restructuring meetings. In many companies, meetings are data-heavy. Executives spend hours reviewing slides and debating metrics. With AI in place, those tasks shrink. Data is processed instantly. Meetings should shift toward interpretation and communication. For example:

- Begin with AI recommendations displayed clearly.

- Spend time discussing how to explain the decision to employees.

- Identify which stakeholders need personal engagement and who delivers it.

- Create communication plans alongside strategy plans.

This change ensures meetings reflect the new balance between AI precision and human interpretation.

Executives should also build "story units" in their organizations. These are teams trained to translate AI-driven strategies into narratives for employees, investors, and customers. They combine communication specialists with leaders who understand the strategy. The role is not marketing spin. It is meaning-making. It connects numbers to purpose.

Boards must also adapt. Oversight shifts from evaluating whether executives made the right call to whether executives can communicate and uphold the AI-driven call with integrity. That requires new performance metrics. Boards should assess leaders not just on profitability but on trust-building, communication, and cultural alignment. They should ask: how well did this leader explain the AI's recommendation? How did employees respond? Did trust rise or fall?

The scholarly debate here is active. Some argue that as AI improves, it will eventually simulate empathy and even craft compelling stories. Natural language models already write speeches and advertising copy. Skeptics argue these outputs are hollow, lacking genuine connection. The truth likely lies in between. AI may assist in creating drafts, but authenticity comes from human presence. A leader speaking with lived conviction is not the same as a script read from a machine. People know the difference. That human spark will remain irreplaceable.

Executives should also prepare for tension. Some will feel their roles reduced as AI takes over core decision-making. The challenge is to reframe. Leadership is not about controlling every number. It is about guiding meaning. It is about being the human face of choices shaped by data. Leaders who embrace this shift will thrive. Those who resist may find themselves sidelined.

Practical steps to reinforce this transition include:

- Establishing executive training programs focused on communication and empathy.

- Creating new metrics for leadership performance that track trust, not just profit.

- Embedding human-led negotiation teams in areas where relationships matter most.

- Building systems for leaders to test and refine their storytelling skills.

- Encouraging leaders to hold direct conversations with employees, not just issue memos.

Another critical role for humans is ethics. AI can process data, but it cannot define right and wrong. Ethical questions arise constantly in corporate life. Should a company exit a profitable market where human rights are at risk? Should it prioritize shareholder returns or environmental sustainability? AI can show trade-offs, but the decision rests on values. Those values must be articulated and defended by human leaders.

Executives must prepare themselves for these moments. They should develop clear ethical frameworks that guide their responses. Boards should expect leaders to articulate values as part of their roles. When AI presents options, leaders should frame the choice in ethical terms and communicate why it matters. Ethics is not a data problem. It is a human responsibility.

The cultural role of humans also expands. Employees don't bond with algorithms. They bond with leaders who inspire and represent them. In times of crisis, they look to people, not machines. Leaders must be present, visible, and engaged. They must show empathy when jobs are cut, passion when new projects are launched, and humility when mistakes are made. These gestures build culture, which remains beyond the reach of AI.

The future of executive leadership will therefore be hybrid. AI will handle integration, scale, and bias-free decision-making. Humans will handle meaning, empathy, negotiation, vision, and ethics. This division of labor is not a weakness. It is strength. It recognizes that leadership is more than numbers. It is about connection.

Executives who want to prepare should ask themselves hard questions. How strong are my storytelling skills? Do I connect with employees on an emotional level? Can I negotiate not just with numbers but with pride and dignity? Do I hold a vision that inspires, even when data suggests caution? Am I prepared to defend ethical choices when they conflict with profit? These questions define the human edge in the age of AI.

The rise of AI does not eliminate humans from leadership. It redefines them. Those who embrace empathy, communication, vision, and ethics will remain invaluable. Those who cling to control of numbers will be replaced. The future belongs to leaders who understand that humans and machines are not competitors but complements — each powerful in different ways, together stronger than either could be alone.

9: The Boardroom Without a Corner Office

The corner office has always symbolized power. It carries the weight of authority, the aura of prestige, and the assumption that one person sits at the top, directing the enterprise. For generations, boards of directors were designed to oversee that individual, to check their judgment, to approve their pay, and to step in when their performance faltered. But when leadership shifts from a person to a system, the structure of governance itself must change. What does oversight look like when there is no human CEO to supervise? What does accountability mean when decision-making flows from algorithms rather than a personality? These are no longer theoretical questions. They are becoming real challenges as corporations experiment with AI-driven leadership.

Boards today carry three main responsibilities: setting direction, monitoring performance, and holding management accountable. These functions assume a management team led by a human CEO. The relationship is personal. Directors ask questions, probe for weaknesses, and judge character as much as results. AI leadership breaks that model. Systems do not attend dinners with board members, charm investors, or shift tone depending on the room. They do not negotiate their own pay. They operate continuously, offering outputs rather than conversation. This forces boards to redefine their role.

One of the first shifts is moving from personality oversight to system oversight. With human CEOs, boards often judge leadership by style, charisma, or trustworthiness. With AI, boards must judge by performance data. Did the system's decisions meet strategic objectives? Were the recommendations accurate over time? Did risk models reflect reality? Oversight becomes less about evaluating character and more about auditing algorithms. This requires new expertise on boards. Directors will need literacy in data science, model validation, and algorithmic transparency. Without it, they cannot fulfill their fiduciary duty.

The governance challenge is not only technical. It is structural. Traditional boards assume a hierarchy where the CEO reports up. But an AI system is not a reporting executive. It is a decision hub integrated across finance, operations, HR, and markets. Oversight cannot rely on a single point of contact. It must be distributed, layered, and continuous. Directors will need to establish permanent committees to monitor system design, data inputs, and ethical safeguards. This is closer to how audit committees review financial controls than how boards currently evaluate a CEO.

Consider the role of accountability. When a human CEO makes a poor decision, boards can fire them. Accountability is clear. With AI, responsibility is diffuse. If the system underperforms, is the blame with the vendor who built it, the engineers who maintained it, or the board that approved it? Clear accountability frameworks must be defined before adoption. Boards should create contracts that specify ownership of system risks, much like liability contracts for products. Without this, companies face legal and reputational exposure.

Implementation requires concrete steps. Boards preparing for AI leadership should:

- Establish an AI oversight committee, staffed with directors who have technical literacy.

- Require regular audits of AI models, just as financial audits are required by law.

- Define clear accountability between vendors, internal teams, and the board for system errors.

- Set thresholds for human override, identifying which types of decisions require board sign-off.

- Develop reporting systems that allow AI decisions to be explained in human terms to directors.

The last point is critical. Explainability sits at the center of governance. Boards cannot oversee what they cannot understand. If an AI system recommends divesting from a market, directors must be able to see the reasoning chain: demand projections, regulatory risk, cost models. Without transparency, oversight collapses. Boards must demand explainable AI, not just black-box outputs.

The scholarly debate is divided. Some argue that AI-driven leadership reduces the need for boards altogether. If systems can optimize strategy continuously, why not reduce governance structures to legal compliance only? Others argue the opposite: that AI requires more governance, not less, because risks are harder to detect and accountability is diffuse. The emerging consensus is that boards will not disappear but will be forced to reconfigure. They must act less as overseers of personalities and more as guardians of process, design, and integrity.

Culture also shifts in this model. Boards often act as a counterweight to executive ego. They temper ambition, rein in excess, and question assumptions. When AI replaces the CEO, ego is gone. But new risks emerge: blind faith in system accuracy, or complacency in assuming the model cannot fail. Boards must build cultures of skepticism, not of distrust but of inquiry. They must ask hard questions: Which datasets shape this recommendation? What assumptions are embedded in the model? What risks are being overlooked because the algorithm has never seen them? Skepticism must be institutionalized.

Boards will also need to reshape their cadence. Today, most boards meet quarterly. AI systems, by contrast, run continuously. A quarterly meeting cycle is mismatched to real-time decision-making. Directors may need to operate in rolling sessions, supported by dashboards that provide live updates. They will need to shift from episodic oversight to continuous monitoring. This doesn't mean directors work full time, but it means oversight is augmented by systems that alert them when thresholds are crossed. Governance becomes dynamic.

This evolution also raises questions of composition. Traditional boards often include industry veterans, legal experts, and financiers. With AI leadership, boards will require technologists, ethicists, and data governance specialists. Diversity of expertise becomes as critical as diversity of demographics. Directors who cannot interpret algorithmic reasoning will be at a disadvantage. Shareholders will demand competence in this new domain.

A related issue is regulatory compliance. Governments are beginning to regulate AI in finance, healthcare, and employment. Boards must ensure compliance with these frameworks. That means not only monitoring AI outcomes but also documenting decision trails for regulators. When AI is the CEO, regulators will ask: who signed off on its actions? Boards must be prepared to answer. That requires robust reporting, audit trails, and clear assignment of accountability.

The financial community is already paying attention. Institutional investors such as BlackRock and Vanguard emphasize governance quality in their portfolio reviews. They will scrutinize how boards oversee AI. Weak governance will be punished with shareholder activism. Strong governance will be rewarded with trust and capital. Boards must prepare for this scrutiny now, not after problems arise.

Executives can prepare their boards for this shift by running pilot programs. Begin by placing AI in charge of narrow domains — capital allocation within a division, supply chain optimization, or HR planning. Create board committees to oversee these pilots. Document the governance process. Learn what questions directors must ask, what expertise they lack, and what reporting systems they require. Use these pilots to build muscle before expanding AI's authority.

Communication is another key task. Boards must not only oversee AI but explain it. Shareholders, employees, and regulators will demand reassurance. Directors must be able to say, with clarity, how AI shapes decisions, where humans remain involved, and how accountability works. Silence or vagueness will breed distrust. Transparency must become a board competency.

This raises another practical need: education. Most directors today are not trained in AI oversight. Boards should establish mandatory training programs, much like they do for compliance and audit. These programs should cover model basics, risks, governance frameworks, and ethical considerations. Without education, directors cannot fulfill their duty of care.

Another structural innovation may be the rise of "AI trustees." These are independent bodies tasked with auditing and certifying AI governance. Just as auditors review financial controls, trustees could review AI systems for fairness, accuracy, and compliance. Boards could rely on these trustees as external checks. This would reassure investors and regulators that oversight is not captured by management or technical teams.

The cultural signal of a boardroom without a corner office is powerful. It suggests that corporations are no longer led by single figures but by collective intelligence — both human and machine. That changes not only governance but identity. Companies will need to rethink how they present themselves to markets and employees. No longer can they highlight a charismatic CEO on the cover of annual reports. They must highlight systems of governance, integrity of data, and clarity of purpose. Boards will be central to shaping this new identity.

Executives who prepare now will position their companies ahead of the curve. Those who wait will face crises of trust, accountability, and compliance when AI leadership becomes the norm. The corner office may disappear, but governance cannot. The boardroom remains, transformed into the guardian of systems rather than the supervisor of personalities.

"Boards will no longer ask who leads, but how the system decides."
— *Dave Karpinsky*

Part IV: Embracing the Shift

10: The Early Adopters: Companies Leading the Way

The business press often celebrates visionaries — the CEOs who seem larger than life. Their names dominate headlines, their words move markets, and their images fill magazine covers. But beneath the surface of this familiar narrative, something less visible is happening. In boardrooms and operations centers, in marketing departments and logistics hubs, companies are quietly testing a different kind of leader. They are not waiting for the debate about AI in executive roles to resolve. They are already experimenting with it.

These early adopters don't advertise their moves with splashy campaigns. They rarely frame it as "AI taking over leadership." Instead, they describe it as advanced analytics, intelligent automation, or strategic decision support. Yet if you follow the trail of outcomes, it becomes clear. They are ceding executive functions — capital allocation, pricing, workforce planning, supply chain strategy — to systems that don't tire, don't negotiate, and don't carry personal bias. For boards and executives watching closely, these experiments offer a preview of what a company without a traditional CEO might look like.

Finance has been one of the most visible testing grounds. Hedge funds and asset managers have long experimented with algorithmic decision-making. Renaissance Technologies built its reputation by letting models, not humans, drive trades. BlackRock introduced its Aladdin system to evaluate risk and allocate assets across trillions under management. These platforms are more than tools. They actively determine where capital flows. In traditional firms, such calls would sit at the CEO or CIO level. Here, the authority sits with algorithms. Boards still oversee, but the system does the heavy lifting.

Insurance provides another window. Large insurers like AIG and Allianz now rely on AI platforms to price policies and assess risk across geographies. These aren't just back-office tools. They influence which markets the company enters, which lines of business expand, and which are shut down. That's strategic decision-making at the highest level. By letting systems lead, these companies reduce exposure to human misjudgment while increasing precision. It's governance, underwriting, and growth strategy all rolled into code.

Supply chain management has also become a proving ground. DHL and UPS operate networks so complex that no human executive could oversee them in real time. Their AI-driven logistics systems weigh weather disruptions, fuel costs, geopolitical risks, and demand surges. The systems reroute shipments, adjust staffing, and shift resources instantly. In effect, they run the global nervous system of these companies. Executives still set long-term direction, but day-to-day authority has shifted to machines. This is not support — it is leadership at operational scale.

Retail giants are also experimenting with AI-driven leadership. Walmart and Target use AI systems to decide pricing, inventory levels, and promotions across thousands of stores. These decisions, once debated in long meetings, are now automated at scale. The financial impact is massive. A slight shift in pricing across millions of SKUs can add billions in profit. Boards may still hold the CEO accountable for strategy, but in practice the strategy is executed — and often determined — by AI systems that learn and adapt faster than any human committee.

Marketing has been transformed in similar ways. Coca-Cola, for instance, uses AI platforms to design and test ad campaigns, deciding which messages resonate in which markets. Netflix uses algorithms to determine what content to produce, how to price subscriptions, and which shows to promote. These choices are not minor. They define the brand, shape the customer relationship, and drive billions in revenue. AI has moved far beyond targeting ads. It now makes executive-level calls on where companies place their bets.

The question for executives isn't whether this is happening — it already is. The question is how to prepare their own organizations to adopt similar systems. Early adopters reveal some practical steps.

The first is defining the scope. No company hands its entire strategy to AI overnight. They begin with domains where data is rich, outcomes are measurable, and human bias creates inefficiency. Finance, logistics, and marketing meet those conditions. From there, the scope expands. Executives who want to follow suit should map their own organizations. Where is data abundant? Where are decisions frequent, measurable, and prone to politics? Start there.

The second step is building trust. Adoption falters when employees or directors see AI as a black box. Early adopters focus on explainability. BlackRock's Aladdin platform doesn't just spit out recommendations. It explains risk exposure, correlations, and projected outcomes. UPS's logistics systems show why routes shift, not just that they did. Transparency turns skepticism into trust. Executives should demand that their AI systems present reasoning, not just results.

Another lesson from early adopters is integration. AI doesn't work well when bolted onto silos. It thrives when data streams converge. Walmart's pricing systems integrate customer purchase history, supplier costs, weather forecasts, and competitor pricing in one hub. That integration makes its recommendations stronger than any isolated analysis. Executives who want to build AI leadership must first break down silos. That means aligning finance, operations, HR, and marketing data on common platforms. Integration is not just a technical task. It is a cultural one.

Boards play a critical role here. Early adopters often created new committees to oversee AI systems. These committees audit models, monitor outcomes, and assess risks. They don't evaluate a human CEO. They evaluate performance data. Boards that want to prepare should do the same. Establish governance frameworks that treat AI systems as accountable entities. Define who owns risk. Require regular audits. Create escalation paths for when systems underperform. This transforms oversight from personality-driven to performance-driven.

Cultural adaptation is another visible lesson. Employees often resist when they feel decisions are made by machines. Early adopters tackled this by pairing AI authority with human communication. At UPS, route changes driven by algorithms are explained by managers to drivers. At Netflix, content decisions shaped by algorithms are communicated to creative teams in terms of audience demand. The system may make the call, but humans still deliver the message. That combination preserves trust. Executives must recognize that AI leadership requires human storytelling. Numbers alone won't secure buy-in.

Scholarly debates mirror these lessons. Some argue early adopters risk over-reliance on AI, creating blind spots when systems misinterpret data. Others argue the opposite — that failing to adopt creates competitive disadvantage. The truth lies in how governance, transparency, and human roles are defined. Early adopters show that risk can be managed. What cannot be managed is falling behind competitors who adopt first.

Executives who want to implement AI-driven leadership can start with pilot projects. Identify one domain with rich data and measurable outcomes — pricing, supply chain, or workforce planning. Deploy AI with clear metrics. Monitor outcomes against human-led baselines. Document gains and failures. Use these pilots to build board trust, employee comfort, and system refinement. Expansion can follow.

Here are some practical activities to consider:

- Run controlled trials: Assign AI to set pricing in one region while humans set pricing in another. Compare results.

- Establish AI accountability logs: Record every recommendation, the human response, and the outcome. Review patterns quarterly.

- Design communication protocols: Decide how managers explain AI-driven decisions to employees and customers.

- Audit vendor relationships: Ensure contracts specify responsibility for AI performance and errors.

- Integrate scenario planning: Use AI to run millions of simulations for capital allocation, supply chain disruptions, or product launches. Present these to boards as part of strategic discussions.

The early adopters also highlight a broader shift. Authority is no longer tied to titles alone. In many companies, the real decision-maker is already a system. The human CEO may still sign reports, attend conferences, and speak with investors. But the calls that drive profit, shape markets, and define direction are made elsewhere. Boards and shareholders may not always admit it, but they know. Authority has shifted.

Executives must decide whether to resist this shift or harness it. Resisting means clinging to structures where human leaders hold nominal authority while systems quietly decide. Harnessing means being intentional: defining governance, building transparency, and reshaping roles. Early adopters are proving that the latter approach creates advantage. It reduces cost, increases speed, and improves accuracy.

The cultural narrative will take longer to catch up. Journalists will still focus on celebrity CEOs. Investors will still look for a human face to rally behind. But within the enterprise, authority is already migrating. It is not a future scenario. It is the present reality for companies willing to act.

Executives reading this should reflect on their own organizations. Where are decisions already more data-driven than human-driven? Where do systems quietly set direction without public acknowledgment? What would change if those systems were brought into the open, treated as leadership, and overseen accordingly? The early adopters show that this is possible. They also show that it is profitable.

The lesson is clear. Waiting for certainty means falling behind. Acting now means shaping the future. The companies leading the way are not the loudest. They are the ones willing to shift authority, restructure governance, and redefine leadership before anyone else admits the corner office is no longer at the center of decision-making.

11: From Skepticism to Strategy

The first time a board hears the phrase "AI leadership," the room rarely fills with applause. More often, it fills with doubt. Directors ask who is accountable if something fails. Executives wonder whether their judgment is being replaced. Shareholders worry about reputational risk, and regulators raise questions about compliance. The skepticism is not irrational. For decades, leadership has been tied to human identity — face, voice, judgment, and accountability. Replacing that with algorithms feels disruptive to the very idea of corporate life. Yet history shows that skepticism often precedes adoption. Railroads were once dismissed as unsafe. Computers were dismissed as toys. Cloud computing was dismissed as insecure. Each of those doubts gave way to strategy when the benefits became undeniable.

Resistance to AI leadership takes many forms. Some is rooted in fear of the unknown. Some in protection of personal power. Some in genuine concern about governance and ethics. For AI to move from theory to practice, companies must address each form of skepticism directly, not dismiss it as ignorance. Skeptics often raise the very issues that, when resolved, build stronger systems.

Executive resistance is perhaps the most visible. Many leaders view AI as a threat to their role. For them, authority is tied to decision-making. If algorithms take over, what remains? The key is reframing. Leadership is not disappearing. It is shifting. Executives retain roles in storytelling, empathy, vision, and ethics — the areas AI cannot reach. The task is to make this shift explicit. Organizations must design new leadership development programs that emphasize human skills rather than technical oversight. By showing executives where they remain essential, companies reduce fear and turn resistance into engagement.

Boards must also tackle the power question head-on. When systems take over decisions, boards may ask: what happens to accountability? A human CEO can be hired or fired. An AI cannot. Skepticism here is structural, not emotional. The answer lies in governance frameworks. Boards should build oversight structures that treat AI as a system to be audited, validated, and monitored. This includes creating AI oversight committees, requiring independent audits, and defining escalation protocols. Accountability shifts from personality to process. Making this shift clear builds trust with directors.

Shareholder skepticism often centers on risk. Investors fear being associated with scandals tied to algorithmic bias, opaque decisions, or system failures. They need assurance that adoption will not create reputational damage. Companies can address this by communicating not only the financial benefits but also the safeguards. Publish transparency reports showing how AI systems are tested, validated, and monitored. Explain how human oversight remains in place. Shareholders reward companies that balance innovation with responsibility. Opacity breeds resistance. Transparency turns caution into support.

Regulators present another barrier. Their skepticism is often sharper because their role is to protect the public. Regulators in finance worry about systemic risk if algorithms drive markets unchecked. Regulators in healthcare worry about patient safety. Regulators in employment worry about bias and discrimination. To overcome this, companies must engage early. Waiting until after adoption invites penalties and restrictions. Early adopters in logistics and finance learned that by partnering with regulators, they could shape the rules rather than fight them. Engagement builds credibility and allows companies to set standards that work for both innovation and compliance.

Organizations can take concrete steps to turn skepticism into strategy. The first is mapping the sources of resistance. Create a matrix that identifies where skepticism resides — executives, boards, shareholders, regulators — and what concerns drive it. For each group, design targeted responses. For executives, emphasize role redefinition. For boards, emphasize governance frameworks. For shareholders, emphasize transparency. For regulators, emphasize engagement. Addressing each constituency with precision avoids blanket reassurances that satisfy no one.

The second step is building pilot programs that prove value without high risk. Instead of handing the entire company over to AI, assign specific domains. For example, allow AI to set regional pricing, allocate logistics resources, or manage workforce scheduling. Track outcomes rigorously. Compare them to human-led baselines. Share results with boards, investors, and regulators. Pilots reduce fear by showing evidence. They move the conversation from "what if" to "what works."

Communication is another pillar. Too often, companies adopt AI quietly, fearing backlash. That silence feeds skepticism. The more people feel excluded, the more they resist. Transparency flips the dynamic. Share adoption stories with employees, investors, and regulators. Show the reasoning, the safeguards, and the results. Build trust by making AI leadership visible, not hidden. Silence creates suspicion. Transparency builds legitimacy.

Education is equally important. Most directors, regulators, and investors are not experts in AI. Their skepticism often stems from lack of understanding. Companies should offer structured education programs. Boards should receive training on model validation, ethical risks, and governance frameworks. Investors should be invited to sessions on transparency and accountability. Regulators should be shown demonstrations of systems in action. Education reduces fear by replacing mystery with clarity.

The scholarly debate on skepticism highlights a key point: resistance is not always bad. It can be a form of risk management. Skeptics force organizations to test, validate, and strengthen systems. The danger lies in ignoring skepticism or dismissing it as obstruction. Companies that treat skeptics as partners, not enemies, build more durable strategies. Resistance becomes an asset when it is channeled into accountability.

Executives looking to implement this shift can use a structured approach.

- Listen first: Map where resistance arises and document specific concerns.

- Engage openly: Create forums where skeptics can question and challenge adoption plans.

- Pilot carefully: Use controlled trials to prove outcomes before scaling.

- Report transparently: Publish clear data on performance, oversight, and accountability.

- Educate continuously: Build literacy among directors, investors, and regulators.

- Adapt governance: Create structures that shift accountability from personalities to processes.

The payoff for doing this well is more than adoption. It is competitive advantage. Early adopters who address skepticism effectively face less resistance, build more trust, and expand AI leadership faster. Companies that ignore it stumble, facing backlash from directors, investors, or regulators who feel excluded.

History suggests that resistance is strongest before adoption and fades after results become visible. The first firms to adopt cloud computing faced heavy skepticism over security. Today, no serious company questions its place in enterprise IT. The same trajectory will hold for AI leadership. The skeptics of today will become the advocates of tomorrow once value is proven and risks are managed.

The path from skepticism to strategy is not easy. It requires patience, transparency, and rigor. But it is possible. Companies that embrace the challenge will not only overcome resistance. They will set the standard for how AI leadership enters the mainstream.

12: Re-Skilling the Workforce for an AI-First Future

The shift to AI-led corporations is not only a leadership story. It is a workforce story. As decision-making migrates from the corner office to algorithms, the skills required at every level of the enterprise change. Employees who once relied on instinct must learn to interpret data. Managers who once controlled processes must learn to supervise systems. Teams that once competed for executive attention must learn to collaborate with models that don't tire, flatter, or negotiate. This transformation cannot be managed with slogans or one-off training programs. It requires a deliberate plan to reskill and upskill the workforce for an AI-first future.

Skeptics often frame AI adoption as a zero-sum equation: machines win, humans lose. The reality is more complex. Jobs are not simply eliminated. They are redefined. In logistics, routing decisions once made by dispatchers are now handled by algorithms. Yet dispatchers haven't disappeared. Their roles have shifted toward exception management—handling the edge cases where systems fail or where human judgment is still needed. In finance, analysts who once built spreadsheets now interpret machine forecasts, advising leaders on implications rather than building raw models. These shifts require new skills, not fewer people. The companies that thrive will be those that move quickly to reskill their people rather than treat them as expendable.

The first step is recognizing which skills decline in value and which rise. Repetitive, rules-based tasks are increasingly automated. Pattern recognition at scale is handled better by machines. What rises in value are skills that machines cannot replicate: interpretation, creativity, empathy, ethics, and communication. Employees need to move away from memorizing processes and toward building judgment, insight, and adaptability. That requires a redesign of training, not just an expansion of it.

Consider the manufacturing floor. Workers once trained to follow narrow procedures now must learn how to oversee AI-driven systems. When a predictive maintenance system flags a potential equipment failure, the worker must decide whether to trust the model or override it. That judgment requires understanding both the technology and the physical machinery. Training must cover both domains. The lesson is clear: reskilling is not about teaching employees to code. It is about teaching them to think alongside machines.

Executives should approach this shift with structured planning. Start by mapping the skill base of the current workforce. What tasks are already automated? Which ones will soon be? Which ones require human interpretation? Create a skills inventory. From there, design pathways for employees to move from declining tasks to rising ones. Don't assume employees can make that leap alone. Provide training, mentoring, and opportunities to practice new roles.

Practical reskilling strategies include:

- Data literacy programs: Every employee should understand the basics of how AI systems generate outputs. This doesn't mean deep coding. It means being able to question, interpret, and apply insights responsibly.

- Human skills training: Invest in communication, empathy, negotiation, and storytelling. As AI takes over technical tasks, these skills become the foundation of human value.

- Cross-domain exposure: Encourage employees to learn outside their narrow roles. A finance professional should understand operations data. A supply chain manager should understand marketing dynamics. This cross-training builds the flexibility needed when AI integrates across domains.

- Ethics education: Employees should understand the risks of bias, privacy, and accountability in AI decisions. They will be the first line of defense when issues arise.

The challenge for many executives is scale. How do you reskill thousands of employees at once? The answer lies in modular learning. Companies like AT&T and IBM have already shown the value of building learning platforms that allow employees to take short, targeted courses on emerging skills. Executives should create corporate "universities" that focus on AI collaboration skills, updated constantly as technology advances. Training cannot be static. It must be continuous, embedded into career paths rather than tacked on as an afterthought.

Another practical action is to tie reskilling directly to performance management. If training is optional, participation will lag. Companies should link completion of reskilling modules to promotion eligibility, bonus structures, and career progression. This signals that AI skills are not peripheral. They are central to success. When employees see a direct connection between training and advancement, participation rises.

Cultural support matters as much as structure. Employees often fear reskilling because they associate it with redundancy. They assume that if they need to retrain, their old jobs are vanishing. Leaders must frame reskilling differently. It is not a threat. It is an investment. Executives should communicate clearly: "We are investing in your skills so you remain valuable in this new environment." Framing reskilling as empowerment rather than replacement changes the narrative and increases adoption.

The scholarly debate around reskilling highlights tensions. Some argue automation will create mass unemployment, making retraining a losing battle. Others argue new roles will emerge faster than old ones disappear, making retraining critical. The evidence supports the latter. Historical shifts—from agriculture to industry, from industry to services—always created new roles. The difference today is speed. That makes structured reskilling more urgent than in past transitions.

Executives can implement reskilling through pilot projects. Start with one division. Introduce AI tools, pair them with training, and monitor outcomes. Document what skills employees struggle with, what training formats work, and how managers adapt. Use these pilots to refine the approach before scaling across the enterprise. Treat reskilling as an experiment, not a one-time program.

Companies should also create "AI champions" within teams. These are employees trained deeply in AI collaboration who act as mentors for their peers. Rather than relying solely on central training departments, peer-to-peer learning spreads skills faster and builds trust. Employees often learn best from colleagues who face the same challenges daily.

Here are concrete actions executives can take:

- Establish a skills inventory that maps tasks likely to be automated in the next five years.

- Build modular training programs that employees can complete in short bursts without leaving their roles.

- Create career pathways that show employees how to move from declining tasks into new ones.

- Deploy AI champions in every division to mentor peers and reinforce skills.

- Tie training completion to promotion and bonus structures.

- Communicate clearly and often that reskilling is investment, not redundancy.

Another overlooked element is middle management. These managers are often the bridge between strategy and execution. If they resist, reskilling falters. Companies must train managers not only in new skills but in coaching. Their role shifts from directing processes to guiding employees through transitions. Equip managers with frameworks for mentoring, conflict resolution, and team-level adaptation to AI. Without middle management support, reskilling collapses.

Partnerships also accelerate the process. Companies don't need to build every training module from scratch. Partner with universities, online platforms, and industry groups. Tap into external expertise while tailoring programs to corporate needs. Many universities now offer executive courses on AI ethics, data literacy, and machine collaboration. Bringing these resources in-house reduces cost and increases credibility.

Executives must also think globally. Multinational firms face different reskilling challenges in different regions. In developed markets, employees may already have baseline digital literacy. In emerging markets, training may need to start with fundamentals. Design programs that account for regional variation. Don't assume a single model works everywhere.

Finally, measure outcomes. Too many training programs fail because companies don't track results. Build metrics around:

- Participation rates: Are employees actually completing programs?

- Skill application: Are employees applying new skills in daily tasks?

- Performance impact: Are teams with reskilled employees performing better?

- Employee sentiment: Do employees feel empowered or threatened by training?

Report these metrics to boards and shareholders. Show that reskilling is not just rhetoric but measurable progress. This builds confidence that the company is not only adopting AI but preparing its people for it.

The AI-first future will not wait for organizations to catch up. Those that treat reskilling as a core strategy will move faster, adopt deeper, and gain advantage. Those that treat it as a side project will stumble. The early adopters are already proving this. Their workers are not only adapting to AI — they are thriving because of it. The question is no longer whether reskilling is necessary. It is how quickly and effectively it can be done.

"Charisma may win headlines, but transparency will win markets."

— *Dave Karpinsky*

Part V: Risks and Ethical Frontiers

13: The Trust Dilemma

Every leadership transition raises questions of trust. When a new CEO takes charge, boards ask if this person can be trusted to act with integrity, investors ask if they can be trusted to protect returns, and employees ask if they can be trusted to keep promises. Trust has always been the currency of leadership. When leadership shifts from people to algorithms, that currency is tested in ways no boardroom has faced before.

AI systems bring precision and scale, but they also bring opacity. They can process billions of data points in seconds, yet the path from input to output often remains hidden. Even the engineers who build them can struggle to explain why a model delivered one recommendation over another. This is what scholars call the "black box problem." For corporations, it becomes a trust dilemma: how do you delegate power to a system you cannot fully explain?

The dilemma is amplified by accountability. When a human CEO fails, the board can fire them, regulators can fine them, and shareholders can vote against them. Responsibility is personal. But when an AI system fails, who is accountable? The vendor that built it? The engineers that maintained it? The board that approved it? Without clear accountability, trust weakens. Shareholders worry. Regulators intervene. Employees hesitate to follow. The promise of AI leadership cannot be realized without solving the trust dilemma.

Transparency is the first step. Companies must demand explainable AI, not just accurate AI. Accuracy without explanation cannot sustain trust. If a system recommends divesting from a region, directors must see the reasoning chain—declining demand, rising regulation, geopolitical instability. If a system recommends reallocating capital, executives must understand the drivers—cash flow forecasts, competitor performance, credit market shifts. Without transparency, decisions look arbitrary, and arbitrary decisions erode confidence.

Some executives argue that transparency slows innovation. They claim that as long as models are accurate, explanations don't matter. But this ignores how trust is built. Accuracy can win short-term credibility. Trust requires repeatable, understandable reasoning. Employees are more likely to follow decisions they understand, even when they disagree. Regulators are more likely to allow innovation when they see clear reporting. Transparency is not a barrier. It is the bridge to adoption.

Practical steps are available. Companies can:

- Require AI vendors to provide explainability tools that show reasoning chains in human language.

- Establish internal audit teams tasked with reviewing model assumptions and outputs regularly.

- Develop dashboards for boards that translate system outputs into clear summaries, highlighting drivers and risks.

- Publish transparency reports for investors, outlining how AI systems shape decisions and what safeguards exist.

Accountability frameworks must complement transparency. Boards should define in advance who owns system errors. If an algorithmic trading system misallocates billions, the board cannot simply point to engineers. Accountability must be contractually assigned between vendors, management, and directors. Regulators are already moving in this direction.

The European Union is furthest along. The EU AI Act, expected to be fully enforced in 2026, classifies AI systems by risk. Leadership-level AI — those making financial, employment, or health-related decisions — will be deemed "high risk." Companies must document how models are trained, monitor performance, and maintain audit trails. They must also provide "meaningful information" about how outputs are generated. Boards that ignore this will face fines measured as a percentage of global revenue, not just local profits. For executives, this means transparency is not optional. It is compliance.

In the United States, the SEC is stepping into the conversation. While no AI-specific act exists yet, the SEC has made it clear that public companies must disclose material risks tied to AI. If algorithms drive strategy, boards will be expected to show oversight in 10-K filings. The agency is also reviewing whether AI in financial decision-making creates conflicts of interest. For companies, this signals that oversight committees, disclosure processes, and board education are not only best practices but soon regulatory expectations.

China has moved even faster, particularly in algorithm governance. Regulations issued by the Cyberspace

Administration of China require companies to register algorithms with the state, disclose how they rank, recommend, or allocate, and ensure outputs align with "social values." While these rules differ from Western frameworks, they highlight a global trend: governments are not waiting. They are demanding oversight, transparency, and accountability now. Executives of multinationals must prepare for fragmented compliance regimes where the same AI system may face different reporting obligations in each region.

Another way to strengthen trust is by defining thresholds for human override. No board will delegate every decision to machines. Certain categories — mass layoffs, market exits, acquisitions — require human sign-off, even if systems recommend them. Setting these thresholds in advance clarifies accountability. It reassures employees, shareholders, and regulators that machines are not operating unchecked. It also reinforces the human role in ethics and judgment, areas AI cannot address.

The scholarly debate on trust in AI highlights a central tension. Some argue that as systems prove themselves over time, trust will build naturally. Just as people trust planes without understanding aerodynamics, they may trust algorithms without understanding code. Others argue that leadership is different. Planes operate in physical systems governed by stable laws. Markets, politics, and human behavior are less predictable. Blind trust in algorithms can create systemic risks. The balance lies in building systems that are both accurate and explainable, with clear lines of accountability.

For executives, building trust in AI leadership requires action on multiple fronts. Start with education. Boards and senior leaders must be trained to interpret algorithmic outputs. Without literacy, they cannot exercise oversight. Partner with universities or consulting firms to design training modules that cover model basics, explainability tools, and ethical risks. Literacy builds confidence, and confidence builds trust.

Second, invest in governance. Create AI oversight committees that function like audit committees. Their role is not to manage day-to-day operations but to monitor system integrity. They should review reports, question assumptions, and escalate concerns. By institutionalizing oversight, companies reassure stakeholders that accountability exists.

Third, prioritize communication. Shareholders and employees must understand how AI shapes decisions. Too often, companies adopt systems quietly, fearing backlash. That secrecy erodes trust. Transparency, even about limitations, builds credibility. Publish clear communication explaining which decisions are AI-driven, how oversight works, and where humans remain involved. Silence creates suspicion. Openness builds legitimacy.

Practical examples reinforce this point. In finance, BlackRock's Aladdin system generates recommendations on capital allocation. The company communicates openly about how it uses Aladdin, explaining both benefits and limits. In logistics, UPS's routing system shows drivers not only new routes but also the reasons — traffic, weather, fuel costs. These explanations turn resistance into compliance. Trust is not automatic. It is earned through clarity.

Companies should also plan for failure. AI systems will make mistakes. The question is not if but when. Trust depends on how companies respond. If boards deny responsibility, trust collapses. If they take ownership, explain the failure, and show corrective action, trust strengthens. Just as human leaders build credibility by admitting mistakes, companies build credibility by owning AI errors. Planning response protocols in advance ensures consistency and speed when failures occur.

The trust dilemma also involves ethics. AI systems may recommend profitable actions that conflict with corporate values. For instance, an AI might recommend entering a market with weak labor protections. The numbers look good. The ethics do not. Trust requires boards to step in, frame the choice ethically, and override when necessary. Ethics cannot be delegated to machines. They must be owned by humans. Companies that fail here will face public backlash, eroding trust not only in AI but in the enterprise itself.

To implement ethical oversight, executives should:

- Establish corporate values as constraints within AI models.

- Create escalation protocols for decisions with ethical implications.

- Require boards to review and sign off on high-risk ethical decisions.

- Train managers to identify when system outputs conflict with company values.

Another dimension of trust is cultural. Employees must trust that AI will not be used against them. If they fear systems will eliminate jobs indiscriminately or monitor them unfairly, resistance will rise. Leaders must communicate clearly how AI will be used, what protections exist, and how employees will be supported through reskilling. Trust at the workforce level is as critical as trust at the board level. Without it, adoption stalls.

Executives should also recognize that trust is competitive. Companies that build transparent, accountable AI leadership will attract investors, employees, and regulators. Those that adopt secretly or irresponsibly will face backlash. Trust becomes a differentiator. Just as companies once competed on brand reputation, they will now compete on AI trust. The market will reward those who get it right.

Building trust requires concrete infrastructure. Companies should invest in:

- Model documentation systems that record data sources, assumptions, and outputs.

- Audit trails that allow regulators and boards to trace decisions.

- Simulation environments where directors can test model behavior under different scenarios.

- Communication channels that explain AI outputs to employees and investors in plain language.

These are not optional features. They are the foundation of trust in AI leadership.

14: Bias in the Machine

Trusting an algorithm to guide a corporation is only as safe as the data that feeds it. Every executive knows the phrase "garbage in, garbage out." Yet when applied to AI, the problem is far more dangerous. Data is not neutral. It carries the fingerprints of human history—who collected it, how it was framed, and what blind spots were ignored. When an AI system trained on biased inputs begins making executive decisions, the risk is not one flawed judgment. The risk is scaling flawed judgments across thousands of employees, millions of customers, and billions of dollars.

Bias in AI is not an abstract problem. It has been documented repeatedly. Recruiting systems trained on historical hiring data learned to prefer male candidates because past executives were overwhelmingly men. Credit scoring models trained on incomplete financial histories downgraded minority borrowers. Healthcare algorithms underestimated needs of certain patient groups because training sets relied on cost of care rather than actual medical outcomes. These examples don't live in the corner of research papers. They show up in boardrooms, lawsuits, and regulatory hearings. For companies shifting leadership power to AI, bias becomes more than a compliance risk. It becomes a reputational and strategic threat.

Bias takes many forms. Sampling bias occurs when training data doesn't represent the population it will be applied to. Historical bias appears when past inequities are baked into the dataset. Measurement bias arises when the wrong proxy is used for success—for example, assuming medical costs equal medical needs. Confirmation bias emerges when developers unconsciously reinforce assumptions in how models are built. An executive doesn't need to master technical detail, but they do need to understand the categories. Without that awareness, bias slips into strategy unchecked.

Why does this matter so deeply for AI-led corporations? Because algorithms don't just reflect data. They amplify it. A biased hiring decision by a human might affect a few dozen candidates. A biased algorithm screening résumés could influence hundreds of thousands annually. A biased credit officer at a bank may harm a community. A biased model allocating loans could reinforce inequality across entire markets. The scale is what turns bias from an ethical challenge into an existential one for companies.

Executives must approach bias with the same rigor they apply to financial risk. The first step is recognition. Denial is the most dangerous position. Leaders must accept that all models carry bias. The task is not elimination—an impossible goal—but detection, mitigation, and governance. That shift in mindset allows boards and managers to build processes that reduce harm rather than pretend perfection.

Concrete steps can be taken. Companies should begin with bias audits. Independent reviews of training data and model outputs can reveal disparities before deployment. Audits should be repeated regularly, not as a one-time event. Bias does not disappear after launch. Data changes.

Markets shift. Models drift. Regular audits ensure continued oversight.

The second step is diverse data sourcing. Many biases come from narrow or incomplete datasets. Expanding data sources reduces blind spots. For instance, a logistics company optimizing delivery routes might include not only traffic patterns but also community feedback on safety, environmental impact, and labor input. Broader data brings broader perspective.

Third, invest in explainability tools. Executives cannot rely on engineers alone to detect bias. They need dashboards that surface disparities clearly: which groups are favored or disadvantaged, what variables are driving recommendations, and how outcomes shift under different scenarios. These tools translate technical outputs into business language, empowering boards to question decisions effectively.

Fourth, create bias accountability structures. Just as financial oversight belongs to audit committees, bias oversight should belong to dedicated governance bodies. These groups review audit reports, demand remediation plans, and escalate issues to the board. Accountability must be formalized. Without it, bias becomes invisible until crisis erupts.

Bias is not only technical. It is cultural. Employees embed their assumptions into data collection and model design. If a sales team historically focused on one demographic, their data reflects that bias. If HR tracked narrow measures of performance, models inherit those limits. To address cultural bias, companies must train employees across functions to recognize how their actions feed AI systems. A procurement officer deciding what data to capture is shaping future AI leadership. That awareness needs to be built into job descriptions, training, and incentives.

This raises a deeper question: what does fairness mean in corporate decision-making? Scholars debate whether fairness means treating everyone equally, producing equal outcomes, or correcting historical disadvantages. There is no single answer. But companies cannot ignore the debate. Boards must define fairness explicitly and encode it into model design. If they don't, algorithms will make implicit choices that reflect hidden assumptions. By setting fairness standards in advance, leaders avoid ambiguity when outputs are challenged.

Regulators are beginning to intervene. The European Union's AI Act requires companies to prove that high-risk systems have been tested for bias. In the United States, the Equal Employment Opportunity Commission has signaled it will scrutinize algorithmic hiring practices. China's algorithm governance rules require companies to align outputs with social values, including fairness. These frameworks differ, but the message is consistent: governments are watching. For global corporations, the challenge is managing multiple regimes simultaneously. An AI system considered compliant in one region may fail in another. Boards must demand region-specific bias controls, not one-size-fits-all approaches.

Bias also creates reputational risk. In the era of social media, a single biased output can trigger global backlash. A bank accused of discriminatory lending can face public boycotts within hours. A retailer accused of biased pricing can lose customer trust overnight. Executives must treat bias not only as a technical issue but as a brand issue. Reputation once damaged is difficult to rebuild.

The dangers are already visible in public controversies. Amazon abandoned its AI-driven hiring tool after discovering it systematically downgraded women's résumés. The system had been trained on ten years of applications, mostly from men. Apple Card faced public outrage when customers noticed women receiving lower credit limits than men with identical financial profiles. The issue reached regulators in New York. In U.S. courts, the COMPAS recidivism algorithm, used to predict likelihood of reoffending, was shown to label Black defendants as higher risk more often than white defendants with similar records. None of these systems were designed with bias in mind. All of them scaled bias because it was already embedded in the data. For executives, the lesson is unavoidable: bias will not stay hidden. It will surface in public view, and when it does, the damage multiplies.

Practical actions for leaders include:

- Commission independent bias audits for every high-impact AI system.

- Build explainability dashboards to give boards visibility into model outputs.

- Define fairness standards at the board level and integrate them into system design.

- Train employees across functions on how their data choices affect bias.

- Appoint cross-functional bias committees with authority to demand remediation.

- Adapt systems to regional regulatory requirements on fairness and transparency.

A forward-thinking executive should also ask: how do we balance bias against performance? Eliminating bias completely can reduce efficiency. For instance, correcting bias in loan approvals may increase default rates. Boards must decide what trade-offs align with corporate values. These decisions cannot be left to engineers. They belong at the highest levels of governance.

The danger of ignoring bias is not only external. Internally, biased systems damage morale. Employees who feel discriminated against or sidelined by AI lose trust in leadership. They may disengage, resist adoption, or even leak internal documents. Building trust inside the company is as important as building trust outside. Clear communication about how bias is monitored and corrected reassures employees that fairness is not a slogan but a practice.

The debate around bias is also shaping academic thought. Some scholars argue that bias is inevitable because data reflects human society, which is itself biased. Others argue that AI can actually reduce bias compared to humans, who act with emotion, fatigue, and prejudice. Both perspectives carry truth. The real question is whether companies design systems that minimize bias actively or allow it to propagate passively. The answer determines whether AI leadership strengthens or weakens corporate legitimacy.

Executives must approach bias as a strategic issue, not just a compliance box. The companies that lead will not be those with perfect models but those with transparent, accountable, and fair models. Investors are beginning to factor this into decisions. ESG frameworks increasingly include algorithmic fairness. Funds are asking whether companies adopting AI are monitoring for bias. Failing to answer these questions may limit access to capital.

Addressing bias requires investment. Audits, explainability tools, training, and oversight committees cost money. Some leaders will balk. But the cost of ignoring bias is higher: lawsuits, fines, reputational damage, and lost trust. Boards should allocate budgets to bias governance just as they allocate to cybersecurity. The return on investment comes not only in risk reduction but in competitive advantage. Companies seen as fair will attract more customers, employees, and investors.

Bias is the Achilles' heel of AI leadership. It is the factor most likely to trigger backlash, regulation, or rejection. But it is also manageable. By treating it with the seriousness of financial or legal risk, boards can transform bias from a hidden threat into a managed variable. Transparency, accountability, and cultural awareness are the tools. The question is whether leaders will use them before crisis forces their hand.

15: Who Bears Responsibility?

When a human CEO fails, responsibility is clear. Boards can remove them, regulators can sanction them, and courts can hold them accountable. But when a decision made by an AI system harms customers, disrupts markets, or costs billions, the question grows sharper: who bears responsibility?

This question sits at the center of AI-led corporations. The appeal of AI is speed, consistency, and freedom from ego. Yet the risk is diffuse accountability. If leadership decisions no longer come from a single person but from a system, responsibility must be reframed. Companies that fail to define this early will discover the vacuum of accountability invites litigation, regulatory intervention, and reputational collapse.

Responsibility can be viewed across three dimensions: legal liability, corporate governance, and ethical stewardship. Each demands its own structure, but they cannot be separated. The law determines who is accountable in courts. Governance determines how accountability is structured internally. Ethics determines how stakeholders judge fairness. Together they form the backbone of trust in AI-led corporations.

In law, the debate has already begun. Courts across multiple jurisdictions have confronted AI indirectly. In the United States, cases involving autonomous vehicles have forced courts to assign responsibility between manufacturers, software developers, and drivers. In Europe, regulators have debated whether AI should be given a form of legal personhood. While no system yet has full legal responsibility, precedent shows that liability will not vanish simply because decisions come from an algorithm. It will be assigned — either to corporations, to boards, or to vendors.

For executives, this means contracts matter more than ever. When licensing AI systems, corporations must define liability clearly. If a vendor provides a model that later produces discriminatory results, who pays the fine? If a supply chain AI misallocates resources and causes loss, who compensates shareholders? Contracts must allocate risk between developers, operators, and overseers. Without this, every dispute lands squarely on the corporation, no matter how outsourced the system appears.

Governance must adapt in parallel. Traditional governance structures are built around human leadership. Boards oversee the CEO. Committees oversee specific risks. When AI makes decisions, those structures must be retooled. A company cannot put an AI system in front of a compensation committee or interview it during succession planning. Oversight must shift from individuals to processes.

One practical step is the creation of AI oversight committees at the board level. Just as audit committees monitor financial reporting, AI committees should monitor algorithmic decision-making. Their role includes approving major deployments, reviewing bias audits, demanding transparency, and ensuring compliance with regulations. This formalizes responsibility. Instead of waiting until regulators demand answers, boards can show structured oversight from the start.

At the management level, accountability requires operational controls. Every decision made by AI should leave an auditable trail. This includes data inputs, reasoning chains, and outputs. When regulators or shareholders ask why a decision was made, executives must be able to point to documentation, not just say "the system decided." Auditability transforms diffuse responsibility into traceable accountability.

Ethics adds another layer. Legal and governance structures may protect corporations from fines or lawsuits, but public opinion demands more. Stakeholders want assurance that AI decisions align with corporate values. Responsibility here cannot be outsourced. A company cannot say, "the algorithm was at fault." The public will not accept that excuse. Responsibility must rest with the enterprise, not the machine.

Examples from recent years make this clear. When autonomous vehicles caused accidents, manufacturers faced public backlash even when drivers were at fault. When social media platforms amplified harmful content, CEOs were called before Congress, not engineers. In finance, when algorithmic trading triggered flash crashes, regulators turned to market operators, not coders. The pattern is consistent: the public and regulators hold companies accountable for outcomes, regardless of technical detail.

The global regulatory environment reinforces this point. The European Union's AI Act requires corporations to maintain documentation proving oversight of high-risk systems. It does not assign liability to the AI. It assigns it to the operators and the companies that deploy it. In the United States, the SEC is signaling that companies must disclose material AI risks in filings. That disclosure implies accountability at the board level. In China, regulations mandate algorithm registration and compliance with social values. Responsibility is anchored to the company, not the code. Across jurisdictions, the principle is consistent: responsibility cannot be delegated away from human oversight.

For executives, the practical implications are clear. Responsibility must be embedded through governance architecture. Boards and management should take the following actions:

- Establish AI oversight committees at the board level, with defined authority and responsibilities.

- Create internal audit functions for AI decisions, with regular reporting to directors.

- Require contracts with AI vendors to specify liability and indemnification clauses.

- Document every high-impact AI decision with auditable reasoning trails.

- Define escalation protocols when AI outputs conflict with ethics or compliance.

These steps shift responsibility from diffuse abstraction to concrete accountability.

The scholarly debate about AI responsibility introduces another dimension. Some argue for "algorithmic personhood," granting AI systems legal identity similar to corporations. This would allow systems to bear liability directly. Others warn that such moves would allow corporations to offload responsibility onto systems, creating a moral hazard. If the AI can be sued, boards may reduce oversight. For now, regulators reject personhood, but the debate signals the tension ahead: how much responsibility can be shifted from humans to systems without eroding accountability?

Executives must prepare for shareholder scrutiny as well. Investors are increasingly asking how companies manage AI risk. Proxy advisors may soon evaluate whether boards have adequate AI governance structures. ESG frameworks are expanding to include algorithmic accountability. Companies that cannot demonstrate clear responsibility will face reputational penalties in capital markets. For executives, this is not theoretical. It will affect share price, voting outcomes, and access to capital.

Responsibility also touches insurance markets. Insurers underwriting corporate liability are beginning to assess AI risk. Policies may exclude claims arising from algorithmic errors unless companies demonstrate strong oversight. This introduces financial pressure. Without governance, insurance costs rise or coverage vanishes. Executives who ignore this will discover responsibility costs embedded in premiums and exclusions.

A forward-looking board should approach responsibility through three lenses: prevention, detection, and response.

Prevention includes bias audits, diverse datasets, and oversight committees. Detection includes monitoring outputs and stress-testing systems. Response includes defined escalation protocols and public communication strategies. Responsibility does not end with prevention. It includes how a company reacts when harm occurs.

Consider the communication challenge. If an AI system makes a decision that harms a community, silence is not an option. Executives must explain what happened, take ownership, and show corrective action. The message cannot be "the algorithm failed." The message must be "we failed in oversight, and here is what we are doing to fix it." Responsibility is demonstrated in transparency and response. Companies that try to hide behind technical excuses erode trust permanently.

Cultural differences complicate responsibility further. In Europe, regulators emphasize human dignity and fairness. In the United States, accountability focuses on disclosure and investor protection. In China, responsibility is tied to social stability and state values. Multinational corporations must navigate these differences. A single AI decision may be judged through multiple lenses across jurisdictions. This requires localized governance, not one global template. Boards must adapt frameworks to regional expectations.

Responsibility is also personal for directors. Regulators are increasingly willing to hold board members liable for oversight failures. Under the EU AI Act, directors of companies deploying high-risk systems may face personal liability for negligence in governance. In the United States, directors could be exposed through securities law if they fail to disclose AI risks. Directors must view AI oversight not as optional but as a fiduciary duty. The stakes are not abstract. They are personal.

Employees add another dimension. If workers feel harmed by AI decisions — biased performance evaluations, unfair scheduling, or discriminatory promotions — they will demand accountability. Labor regulators may intervene. Unions may mobilize. Responsibility here requires not just oversight but communication. Workers must understand how AI systems affect their roles and what recourse exists if errors occur. Without that clarity, internal trust erodes, creating resistance to adoption.

The record of corporate controversies shows how accountability always lands on boards and executives. Boeing's 737 Max crisis highlighted this when software flaws contributed to two fatal crashes. While engineers designed the system, regulators and the public held Boeing's board and executives responsible. Wells Fargo's fake account scandal had algorithmic elements, but accountability landed on senior leaders and directors for oversight failures. Uber's autonomous vehicle accident in Arizona, which killed a pedestrian, drew headlines that named the company and its leaders, not the engineers who coded the perception system. These events reinforce the lesson: in public and regulatory eyes, responsibility flows upward.

The question of responsibility ultimately ties back to legitimacy. Corporations derive legitimacy from accountability. Stakeholders accept decisions, even unpopular ones, when they believe someone is responsible.

If AI decisions appear unaccountable, legitimacy fractures. Shareholders lose faith. Regulators intervene aggressively. Employees disengage. Customers turn away. Responsibility is not just about risk management. It is about sustaining legitimacy in the age of algorithmic leadership.

To implement responsibility effectively, companies should build layered governance:

- Board level: AI oversight committees, integrated into fiduciary duties.

- Executive level: Chief AI officers or equivalents, responsible for operational governance.

- Operational level: Audit teams, compliance functions, and explainability dashboards.

- External layer: Independent audits, regulator engagement, and shareholder communication.

This layered model ensures responsibility is distributed but never diffuse. Each layer has defined duties, each can be audited, and each connects to stakeholders.

The future of responsibility in AI-led corporations will not be settled by technology. It will be settled by governance. Systems will grow smarter. Data will grow richer. But trust will depend on whether boards, executives, regulators, and employees believe responsibility is clear. Without clarity, the promise of AI leadership collapses under the weight of accountability gaps.

"The measure of leadership is shifting from personality to platform integrity."
— *Dave Karpinsky*

Part VI: The New Era of Leadership

16: The Hybrid CEO Model

T he idea of the all-knowing CEO has always been more myth than truth. Even the most celebrated leaders rely on deputies, advisors, and extensive staff networks. No one person can master finance, operations, markets, and culture all at once. The rise of AI forces corporations to confront this reality in sharper terms. The choice is not binary between a human CEO and a machine. The more compelling path lies in a partnership — a hybrid model where humans define vision and values while AI executes with scale, speed, and precision.

This model doesn't just split tasks. It reshapes how leadership itself functions. Instead of a single leader driving decisions through instinct or charisma, companies can design leadership as a collaborative system. The human provides narrative, purpose, and moral grounding. The AI provides analysis, optimization, and execution. Together they create a model that combines what humans do best with what machines do best.

The foundation of the hybrid model rests on acknowledging complementary strengths. Humans excel in context, storytelling, negotiation, and moral reasoning. They can sense nuance in tone, spot contradictions in body language, and interpret silence as meaning. Machines excel in scale, consistency, and pattern recognition. They can process millions of variables simultaneously, learn from real-time data feeds, and forecast outcomes across thousands of scenarios. Marrying these strengths produces a leadership model that is neither nostalgic nor utopian — it is pragmatic.

The hybrid CEO model can be structured across four domains: vision, execution, governance, and communication. Each requires clarity about what is human-led, what is machine-led, and what is shared.

In vision, humans must remain primary. AI can generate scenarios and highlight opportunities, but it cannot replace human intuition about meaning and purpose. A company must articulate why it exists beyond profit. That requires human imagination and empathy. For example, a global apparel company may use AI to model sustainable supply chains, but only human leadership can frame sustainability as a moral obligation and a brand identity. AI cannot define what matters — it can only support how to get there.

In execution, AI dominates. Once the vision is set, AI can allocate resources, optimize supply chains, manage schedules, and adjust in real time. Think of a logistics company balancing fuel costs, driver availability, and customer demand across continents. A human team could meet weekly to update plans. An AI system can do it every hour. The hybrid model frees human leaders from the grind of execution so they can focus on strategy.

Governance sits in between. Both human and machine must collaborate. AI can monitor compliance, scan for anomalies, and flag risks faster than auditors. But humans must decide when trade-offs are acceptable, when reputational risks outweigh efficiency, and when to slow execution for the sake of principle. Governance becomes the space where human judgment filters machine outputs through the lens of values and law.

Communication is entirely human. An AI system can generate reports, draft speeches, and prepare talking points, but trust requires a human face. Employees, investors, and regulators want to hear a voice that feels accountable. An AI may recommend a round of layoffs or a restructuring. But the explanation, the reassurance, and the emotional labor of leading through it must come from a person. This is where leadership returns to its ancient roots: humans connecting with humans.

Practical steps to implement the hybrid CEO model start with role definition. Companies must map which decisions are delegated to AI, which are retained by humans, and which require joint input. Without clear boundaries, accountability fractures. Boards should demand a matrix of decision rights that specifies where AI holds authority and where humans retain veto power.

The second step is transparency. Every AI-driven decision should leave an audit trail that executives can review. When a pricing algorithm adjusts rates across regions, leaders should see what data drove the decision, what scenarios were tested, and what risks were flagged. Transparency prevents blind reliance on the machine and allows humans to step in when context demands.

Third, companies must invest in interfaces that make AI insights usable for leaders. Dashboards that overwhelm with data will fail. What leaders need are tools that highlight trade-offs, visualize outcomes, and allow scenario exploration. The interface is where human and machine meet. If the interface confuses or intimidates, the partnership collapses.

Fourth, training must shift from technical to interpretive. Leaders do not need to code algorithms. They need to understand how algorithms shape decisions, what biases may appear, and how to challenge recommendations constructively. Training programs for executives should include simulations where AI makes decisions and leaders must interpret, question, and refine outputs. This builds fluency without demanding technical mastery.

Fifth, boards should embed hybrid governance structures. An AI oversight committee should work alongside audit and risk committees. Its role is not only compliance but strategic alignment—asking whether the AI's execution supports the company's values and long-term goals. This committee must include directors with technical fluency, but also directors who can probe ethical implications.

The hybrid CEO model is already emerging in pieces. Financial firms use AI to set trading strategies while human executives explain moves to regulators and investors. Airlines use AI to manage scheduling and fuel optimization while human leaders negotiate with unions and governments. Marketing firms use AI to personalize campaigns across millions of customers while human executives frame the brand narrative. The shift is incremental, but the pattern is clear: humans setting direction, AI executing at scale.

Critics may argue that such a model risks hollowing out human leadership. If AI executes everything, do leaders become figureheads? The answer lies in how companies design the role. If leaders abdicate decision-making, they will shrink. But if they claim responsibility for vision, communication, and ethical framing, their role becomes sharper. They no longer drown in operational detail. They lead where only humans can lead.

There are risks. Overreliance on AI may lead to complacency. Leaders may defer too quickly, assuming the machine is always right. This risk must be managed through regular stress tests. Boards should require leaders to occasionally override AI recommendations and document outcomes. This tests both the human and the system, ensuring neither grows unquestioned.

Another risk is conflict between human and machine. What happens when an AI recommends a strategy that conflicts with human values? For instance, an AI might suggest offshoring jobs for efficiency, while human leaders recognize the social cost and reputational backlash. In such cases, the hybrid model must favor human judgment. Companies must codify the principle that AI advises but humans decide when values are at stake.

Global variation adds complexity. In Europe, regulators may demand more human oversight. In the United States, investor pressure may prioritize efficiency. In Asia, state involvement may impose national objectives. Multinational corporations must tailor the hybrid model to local expectations. This means defining decision boundaries differently in different jurisdictions while maintaining a consistent global narrative.

The insurance industry offers a preview of how accountability will shape the hybrid model. Insurers are beginning to write policies that exclude coverage for "autonomous decision errors" unless companies demonstrate governance structures. That means companies must prove they have processes for human oversight, audit trails, and defined responsibilities. Insurers are not just pricing risk. They are enforcing hybrid governance as a condition of coverage.

To make the hybrid CEO model real, executives should take specific actions:

- Map decision rights across vision, execution, governance, and communication.

- Build interfaces that translate AI analysis into business-ready insights.

- Train leaders in interpretive, not technical, skills.

- Establish board-level oversight committees for AI.

- Define escalation rules where humans can override AI outputs.

- Adapt governance for regional regulatory differences.

- Secure insurance coverage by demonstrating robust oversight structures.

Each step moves corporations closer to a future where leadership is not about a single person or a single system but about collaboration.

The cultural impact of the hybrid model will be significant. Employees may initially resist AI in leadership, fearing dehumanization. But when they see human leaders communicating vision, showing empathy, and taking responsibility while AI improves efficiency, resistance can soften. Customers may also adjust quickly if the model delivers consistent value without losing human accountability. Investors, always focused on performance, will likely support it if governance and communication remain strong.

The hybrid CEO model offers not just operational efficiency but a philosophical shift. It acknowledges that leadership is not about knowing everything or doing everything. It is about choosing what to hold onto as distinctly human and what to delegate to systems that can execute better. This clarity is liberating. It allows leaders to focus where they add the most value and to let machines handle what overwhelms human scale.

Boards that embrace this model early will set the standard. They will show regulators that accountability is intact. They will show employees that leadership still has a human face. They will show investors that efficiency and values can coexist. The hybrid model is not a stopgap on the way to full automation. It is likely the dominant model for decades, balancing human purpose with machine execution.

17: The End of Executive Elites

For more than a century, the upper tiers of corporate leadership have resembled a gated community. Access was tightly controlled, granted to those with elite education, powerful networks, or long-standing reputations inside established companies. The corner office signaled privilege as much as responsibility. Corporate success depended on scale, capital, and access to rare executive talent. But what happens when the scarcity of leadership skill is erased? What happens when the intelligence, judgment, and decision-making power once monopolized by a few can be replicated by systems available to many?

The arrival of AI-driven leadership tools is beginning to answer that question. These systems compress decades of executive experience into accessible platforms. They can parse financial data, simulate competitive strategy, manage resource allocation, and forecast outcomes with a speed and depth no human leader can sustain. Just as spreadsheets democratized finance in the 1980s — enabling small firms to perform analysis once limited to big corporations — AI democratizes executive function itself. The implications are profound.

This shift strikes at the heart of corporate inequality. For decades, scale was reinforced by the quality of leadership talent a company could attract. Giants had the means to pay for elite executives with million-dollar packages. Smaller firms often had to make do with leaders who lacked equivalent experience or networks. That gap compounded over time, producing structural inequality across markets. AI changes the equation. When advanced leadership capability becomes a tool, not a person, access is no longer restricted to those who can afford a star CEO.

Consider supply chain management. Multinational corporations spend millions on executives with decades of global logistics experience. They hire teams of consultants to monitor shifting tariffs, track shipping routes, and anticipate geopolitical risks. A small manufacturer could never afford such expertise. Today, AI platforms trained on real-time trade flows, customs data, and commodity markets provide that same foresight at subscription-level cost. A firm with 50 employees can now see the same patterns as a firm with 50,000.

The same holds in finance. Large corporations rely on CFOs supported by armies of analysts. They run scenario planning, manage hedging strategies, and forecast capital allocation. Now AI platforms integrate treasury management, forecasting, and predictive analytics into a single interface. A mid-sized company can access the same caliber of insight without employing dozens of specialists. The playing field levels not through regulation or policy, but through access to machine intelligence.

This does not mean all firms become equal. Scale still matters. Giants have resources to deploy AI at greater depth, to integrate systems across divisions, and to build proprietary models. But the moat shrinks. The differential between the top and the rest narrows. What once required elite pedigree can now be purchased as a service.

To understand the scale of change, think of how cloud computing shifted the economics of IT. Twenty years ago, only global corporations could afford vast data centers. Startups faced crushing infrastructure costs. Cloud computing turned that infrastructure into a subscription, unleashing waves of small competitors able to scale quickly. AI leadership tools operate on the same principle. They transform decision-making power from a scarce asset into a broadly available utility.

For executives, this demands a recalibration of strategy. If your advantage once came from access to rare leadership talent, that advantage is eroding. Competitors once considered beneath notice may soon match your ability to plan, forecast, and execute. The differentiation must shift toward creativity, brand, and customer intimacy — areas where human connection still matters. The structural dominance of executive elites no longer guarantees superiority.

Boards should prepare for this shift by reassessing compensation models. If AI systems can perform much of the analytical and operational work once assigned to highly paid executives, does it make sense to continue paying premium packages? Pressure from shareholders will grow as smaller firms demonstrate similar performance without paying eight-figure salaries. Compensation committees will need new rationales for high pay, focused less on analytical skill and more on leadership presence, narrative power, and stakeholder trust.

The hybrid company of the future may not abolish executive teams, but it will redesign them. Leaders will be valued less for their ability to calculate and more for their ability to interpret, inspire, and persuade. Their partnership with AI will free them from operational complexity, leaving them to focus on vision, culture, and external trust. For firms that once relied on elite credentials alone, this is a humbling transformation.

The democratization of leadership tools carries a social dimension as well. Executive inequality has long mirrored broader economic inequality. The gap between CEO pay and median worker pay widened dramatically in the last fifty years, often justified by the complexity of global decision-making. If AI reduces the scarcity of that skill, pressure will grow to narrow compensation gaps. Workers will question why leaders command disproportionate rewards when much of the technical work is automated. Boards that fail to anticipate this pressure risk reputational harm and regulatory scrutiny.

Practical implementation requires companies to embrace democratized tools at all levels, not just at the top. A hybrid leadership structure works best when managers and directors also have access to AI decision support. This prevents concentration of insight in a few hands and spreads capability across the enterprise. Smaller firms should:

- Adopt AI-driven platforms for finance, supply chain, and HR to match larger rivals.

- Train mid-level managers in interpreting AI insights, not just senior executives.

- Build cross-functional teams where AI outputs are debated collectively, not hoarded at the top.

- Establish governance that ensures transparency and accountability for AI-driven decisions at every level.

Larger firms should:

- Reassess compensation structures tied to analytical skill sets that AI now automates.

- Invest in proprietary data to differentiate their AI tools from off-the-shelf competitors.

- Focus leadership training on communication, negotiation, and narrative skills.

- Expand ethical oversight to maintain trust as automation scales across functions.

The end of executive elites will not erase hierarchy, but it will redefine it. Hierarchy will be justified less by access to knowledge and more by the ability to embody values and connect with stakeholders. A CEO may still hold symbolic power, but the substance of decision-making will be shared with systems that anyone can access. This transition could democratize corporate competition as profoundly as capital markets democratized finance.

History offers parallels. When literacy spread beyond elites, control of knowledge shifted. When printing presses became widespread, authority fragmented. When cloud computing arrived, startups challenged giants. Each time, access to once-scarce capabilities restructured power. AI-driven leadership tools are the next iteration. They promise to erode the mystique of executive elites and broaden the base of competitive capacity.

The debate within management scholarship is already shifting. Traditional theories emphasized resource-based advantages, where firms competed on unique assets like scale, technology, or leadership. Newer theories suggest that advantage will rest less on access and more on adaptability, creativity, and ethical positioning. When every company has comparable analytical capacity, the differentiators move to the human domain — storytelling, trust, and identity.

This shift places new demands on boards. Governance must focus not only on financial oversight but on cultural alignment. Boards will need to assess whether leaders embody values credibly, communicate transparently, and navigate stakeholder trust effectively. The technical skill once prized in the boardroom will be augmented by AI; the human skill of judgment and trust-building will matter more.

Consider global markets. In emerging economies, where access to elite executives was historically limited, democratized tools may produce leapfrogging. Firms in Southeast Asia, Africa, or Latin America can now access AI-driven decision-making equal to Western giants. This could reduce structural inequality across nations as much as across firms. For global corporations, this means competition will not only intensify but diversify. Firms once dismissed as peripheral may quickly become challengers.

Regulators will also play a role. If AI erodes the scarcity of leadership, regulators may intensify scrutiny of executive pay and corporate concentration. Expect debates over whether AI-driven efficiency should translate into lower costs for consumers, higher wages for workers, or simply greater returns for investors. The political implications of democratized leadership will extend beyond corporate walls.

For leaders today, the question is how to prepare. The following steps can position firms for the era of democratized executive power:

- Reevaluate leadership pipelines. Focus on values, narrative skill, and judgment, not just technical expertise.

- Deploy AI leadership tools widely to reduce dependency on a narrow elite.

- Redesign compensation frameworks to reflect the new balance between human and machine.

- Invest in culture, brand, and stakeholder trust as differentiators in a market where analysis is commoditized.

- Anticipate regulatory and social pressure to narrow inequality as leadership scarcity declines.

The end of executive elites will feel disruptive to those who benefited from scarcity. But to most firms, it offers opportunity. Access to world-class decision-making power no longer requires elite pedigrees or massive budgets. What matters is how companies combine machine intelligence with human purpose. In that combination lies the next competitive frontier.

18: The Last CEO

The role of the CEO has always carried a strange duality. It is at once exalted and lonely. Exalted, because the office symbolizes ultimate authority and vision. Lonely, because no one else bears the same burden of accountability. For generations, boards, investors, employees, and governments have looked to a single figure to represent the direction of a corporation. But the very idea of a singular, all-knowing leader was always a fragile construction. In practice, CEOs have leaned on teams, consultants, and networks. They've made decisions based on incomplete data, filtered information, and the constant pull of personal bias. What we are witnessing now is not just the rise of new tools. It is the unraveling of a mythology. The mythology of the singular CEO.

Artificial intelligence changes the balance not by erasing leadership but by distributing it. Instead of a single mind attempting to absorb and synthesize oceans of data, collective intelligence systems integrate finance, operations, markets, customers, and society in real time. Leadership shifts from a person to a process. Authority shifts from personality to platform. The role of humans changes from decision-maker to guide, interpreter, and steward. In this future, the last CEO may not be a person at all but an ecosystem of intelligence.

This shift is already visible in fragments. Risk management systems evaluate thousands of scenarios in seconds. Pricing engines adjust based on real-time demand, weather, or competitor moves. Algorithms track sentiment across millions of social media posts to inform brand strategy. Each of these is a building block in a larger architecture: distributed intelligence applied to leadership.

What makes this moment different from earlier technological changes is the breadth of integration. Past innovations made functions faster — ERP systems improved planning, CRM systems deepened customer knowledge, and BI dashboards enhanced reporting. AI transforms functions into nodes of a shared network. Decisions are no longer isolated in silos but flow across the enterprise like current through a grid. That flow does not pause for quarterly meetings. It updates continuously.

The last CEO is not about replacing human figures with machines. It is about redefining leadership as collective. Humans still matter. They define meaning, values, and the bounds of acceptable trade-offs. They shape the narrative of why the company exists and where it is heading. But the execution, analysis, and adjustment operate at a scale and speed only systems can achieve.

This model of collective intelligence has several defining features.

First, distributed data integration. Companies no longer rely on reports passed upward through hierarchy. Instead, streams of data — from supply chains, sensors, customer interactions, financial transactions, and external signals — feed into shared systems. Decisions emerge from a common pool of intelligence rather than fragmented perspectives.

Second, adaptive execution. Instead of annual plans that quickly fall outdated, strategies update in real time. The system detects shifts in markets, regulations, or social sentiment and adjusts accordingly. Human leaders monitor direction, but the recalibration is continuous.

Third, transparent accountability. Every recommendation and adjustment carries a traceable record. Decisions leave an audit trail of inputs, scenarios, and reasoning. This addresses one of the great weaknesses of human leadership: memory shaped by self-interest. Systems preserve the truth of how decisions were made.

Fourth, shared leadership. No single person embodies the company. Boards, executives, employees, and stakeholders interact with the system as participants. Authority is less about hierarchy and more about dialogue between humans and machines.

The practical implications are enormous. Boards must rethink their role. Instead of evaluating the judgment of one CEO, they must evaluate the design, governance, and performance of a collective intelligence system. Investors must shift from analyzing the personality of leaders to analyzing the transparency, adaptability, and integrity of the systems guiding them. Employees must learn to interact with AI not as tools but as colleagues, questioning, interpreting, and guiding. Regulators must build frameworks for accountability when decision-making is distributed.

For companies, the transition to collective leadership can be managed through a series of deliberate steps.

- Start by identifying critical decision domains — finance, operations, HR, marketing — and mapping how data flows today.

- Integrate systems across domains, building a single intelligence backbone rather than siloed tools.

- Design interfaces that allow human leaders to interpret, question, and guide AI-driven recommendations.

- Establish governance committees focused on AI integrity, bias monitoring, and ethical alignment.

- Train leaders at every level to move from being deciders to interpreters — asking not "what should I decide?" but "what does this recommendation mean in context?"

- Communicate transparently with stakeholders about how AI contributes to decisions and where human judgment remains.

What does this mean for the identity of leadership itself? For more than a century, business schools have celebrated the "great man" theory of leadership — the idea that exceptional individuals shape history through charisma and vision. Collective intelligence challenges that model. Leadership becomes less about charisma and more about curation: curating values, curating meaning, curating trust. The "last CEO" is not a heroic figure but a steward of collective insight.

This shift also changes what employees expect. For decades, workers have seen decisions as opaque, handed down from distant executives. Collective intelligence offers transparency. Employees can see the data and reasoning behind decisions in real time. That can build trust — but only if leaders communicate openly about values and context. Otherwise, transparency without meaning risks alienation.

The cultural challenge is significant. Many organizations still equate authority with hierarchy and personality. Moving to collective leadership requires redefining status. Authority must be measured not by control of information but by ability to interpret it, explain it, and embody its implications. That requires humility from leaders accustomed to prestige and unilateral control. It requires courage from boards to value stewardship over celebrity.

There will be resistance. Some investors will argue that leadership without a face feels risky. Media accustomed to covering charismatic CEOs will struggle to write about collective systems. Politicians may fear loss of accountability. These challenges must be addressed directly. Companies must frame the shift not as the disappearance of leadership but as its maturation — an evolution toward transparency, accountability, and distributed intelligence.

The historical record suggests this resistance will not last. Every major shift in organizational design met skepticism. The move from owner-managed firms to professional managers in the early 20th century was controversial. The rise of boards as independent overseers was seen as diluting authority. The adoption of ERP and enterprise systems faced fierce resistance from executives used to controlling their own spreadsheets. Each time, resistance gave way to adoption once the benefits proved undeniable.

AI-driven collective intelligence offers benefits too powerful to ignore. Speed, consistency, transparency, and adaptability transform how firms compete. Those who cling to personality-driven leadership will find themselves constrained by human limits while rivals scale through systems. The narrative of leadership will change because performance will demand it.

The geopolitical implications are no less significant. Companies in regions with strong AI governance — such as the EU with its AI Act — may gain trust advantages in global markets. Firms in countries with looser oversight may gain speed advantages but face higher reputational risks. The architecture of collective intelligence will not be uniform across nations. But the principle — that leadership shifts from individuals to systems — will be universal.

The social contract of corporations may also be reshaped. If executive pay was justified by the scarcity of decision-making skill, what happens when that scarcity disappears? Collective intelligence exposes the imbalance between executive compensation and median worker pay. Pressure will rise to close gaps. Shareholders may demand more equitable distribution of value when leadership is no longer monopolized by elites. Employees may demand greater voice when they can see decisions unfold transparently. The last CEO could also mean the first step toward democratizing the corporation.

Practical experiments can begin now. Companies do not need to abolish the CEO overnight. They can introduce collective systems incrementally:

- Use AI to manage specific domains such as pricing or logistics, then expand scope.

- Shift board oversight to include system design and ethical monitoring.

- Redesign executive roles to emphasize communication, values, and trust-building.

- Benchmark compensation against contributions that AI cannot replace.

- Create forums where employees engage directly with AI-driven insights to shape local decisions.

Each step builds capacity for a transition from personality-led leadership to collective intelligence.

What then becomes of the CEO title itself? In the near term, it may remain as a symbolic role. Boards, investors, and governments still expect a single point of contact. Over time, though, the function will matter less than the system. The last CEO may be less about abolishing the title and more about rendering it irrelevant. Authority will rest in the integrity of systems, not the charisma of individuals.

The narrative of business history often centers on leaders. Rockefeller, Ford, Welch, Jobs. Future histories may center on systems. The AI backbone that transformed logistics. The collective intelligence that reshaped finance. The governance frameworks that balanced transparency and adaptability. The names of CEOs may fade, replaced by the names of platforms, architectures, and models. That does not diminish the role of humans. It simply relocates it — from being the source of decisions to being the stewards of meaning.

For executives reading this, the challenge is not to resist the change but to shape it. How will you define the role of human leadership when decisions are distributed? How will you frame your company's purpose in a world where intelligence is collective? How will you embody trust when authority shifts from personality to platform? These are not abstract questions. They are the practical design challenges of the next decade.

The last CEO is not a prophecy. It is a design choice. Companies can cling to outdated myths or embrace collective intelligence as the new foundation. Those who choose the latter will discover that leadership is not disappearing. It is expanding — beyond the limits of one mind, into the capacity of many minds, human and machine together.

Epilogue

The story of business has long been told through names. Rockefeller, Ford, Sloan, Welch, Jobs, Musk. Each name became shorthand for a company's identity, strategy, and destiny. The assumption was simple: corporations are reflections of their leaders. Change the leader, change the company. That assumption has carried us through a century of management theory, governance design, and cultural myth. But history has a way of moving past myths. Just as kings gave way to parliaments, and parliaments gave way to democracies, leadership in business is being reshaped.

What this book has argued is not that leadership ends. It is that leadership shifts. Decision-making once concentrated in a single mind is becoming distributed. Analysis once distorted by bias or fatigue is becoming continuous, scalable, and transparent. The executive elite once justified by scarcity is losing its monopoly. In their place, we are seeing the rise of collective intelligence — AI-driven, human-guided, accountable to more stakeholders than ever before.

This transition forces us to reconsider what leadership really means. If it is not charisma or singular vision, then what is it? Leadership becomes the act of curation: curating values, curating meaning, curating trust. Machines can analyze. They can recommend. They can execute. But they cannot believe. They cannot embody purpose. They cannot look another human being in the eye and say, "This matters." That remains our role.

The risk is real. Without governance, AI can scale bias. Without transparency, it can become a black box that erodes trust. Without accountability, it can become a shield for negligence. These dangers are not reasons to resist. They are reasons to design. The companies that thrive will be those that treat AI not as a replacement but as an institution to be governed — subject to the same scrutiny and ethical standards as any other source of power.

And yet, the opportunity is extraordinary. For the first time, the limits of human cognition no longer bind corporate leadership. Every company, regardless of size, can access analytical capacity once reserved for giants. Every board can demand real-time foresight instead of stale reports. Every employee can see the reasoning behind decisions rather than waiting for opaque pronouncements. The center of gravity shifts from the few to the many.

This democratization will change not just companies but markets. Smaller firms will compete with larger ones on strategy, not just survival. Compensation models will shift as executive scarcity declines. Regulators will adapt to systems instead of personalities. And investors will learn to analyze not just the person at the top but the integrity of the platform guiding them.

What comes after the last CEO is not chaos. It is clarity. It is the recognition that no one mind was ever sufficient for the complexity of modern enterprise. The cult of personality may fade, but something stronger will take its place: systems of intelligence guided by shared values and accountable governance. Leadership will not vanish. It will be everywhere.

For today's executives, the question is simple but urgent. How will you prepare? Will you cling to the mythology of singular authority? Or will you design your company for collective intelligence, for transparency, for shared stewardship? The future will not wait. The shift is already underway.

When history looks back on this moment, it may not remember individual CEOs. It may remember the transition. The point where companies chose to redefine leadership not as the power of one but as the wisdom of many — human and machine together. That is the story we are writing now. That is the meaning of the last CEO.

About the Author

Dave Karpinsky, PhD, MBA, PMP, is a globally recognized consultant, executive leader, and professional author whose work bridges business transformation, strategy, and personal development. With over three decades of experience advising Fortune 500 companies, government agencies, and high-growth startups where he traveled to more than 60 countries, Dave brings a rare blend of practical insight, operational excellence, and visionary thinking to every project—and every page.

His career spans top-tier consulting firms including McKinsey & Company, SAP, Accenture, Ernst & Young, Cognizant, Infosys, IBM, and BearingPoint. He has led multi-million-dollar strategic & technology initiatives for global leaders such as Capital One, Coca-Cola, Costco, DHS / TSA, Google, HP, Google, Janus Henderson, John Deere, Lockheed Martin, McLaren, Merck, Nike,, PetSmart, QuidelOrtho, and ViaSat, as well as large-scale public sector programs for the US Government, States of Alaska, Arizona, California, Florida, and Georgia.

As the author of numerous books on project turnaround, leadership, SAP implementation, and personal mastery,

Dave is known for translating complex challenges into actionable strategies that deliver measurable impact. His writing combines analytical precision with compelling storytelling—whether he's decoding enterprise system failures or exploring the psychological dynamics of decision-making and influence.

Dave holds advanced degrees in business, technology and psychology, along with a portfolio of elite professional certifications. He is a sought-after speaker, strategist, and transformation advisor who empowers individuals and organizations to break through barriers and unlock lasting success.

Outside of his professional pursuits, Dave is an avid traveler and photographer, with a passion for astrophotography and a curated collection of high-performance and exotic cars. His global perspective, intellectual curiosity, and relentless drive to improve systems and people continue to inspire readers and clients alike.

To my constant joy and loyal hearts — you make life lighter

Let's Connect

If this book has sparked new ideas, raised thoughtful questions, or inspired fresh strategies in your work or life, I'd love to hear from you. Whether you're looking to exchange insights, explore collaboration, share your own experiences, or continue the conversation, please don't hesitate to reach out. Your feedback and stories help keep these ideas alive and evolving.

You can contact me directly at:
davekarpinsky@yahoo.com

Let's shape the future of enterprise transformation — together.

You can also follow me on:

X: https://x.com/KarpinskyDave
Medium Profile: https://medium.com/@davekarpinsky
Substack: https://davekarpinsky.substack.com
LinkedIn: https://www.linkedin.com/in/dave-karpinsky-mba/

"The last CEO will end the age of personality and begin the age of collective intelligence."

— *Dave Karpinsky*

www.ingramcontent.com/pod-product-compliance
Lightning Source LLC
Chambersburg PA
CBHW071229210326
41597CB00016B/1991